LAUNCH PHOTOGRAPHY

Ben Cooper Photographs Rockets of NASA and More

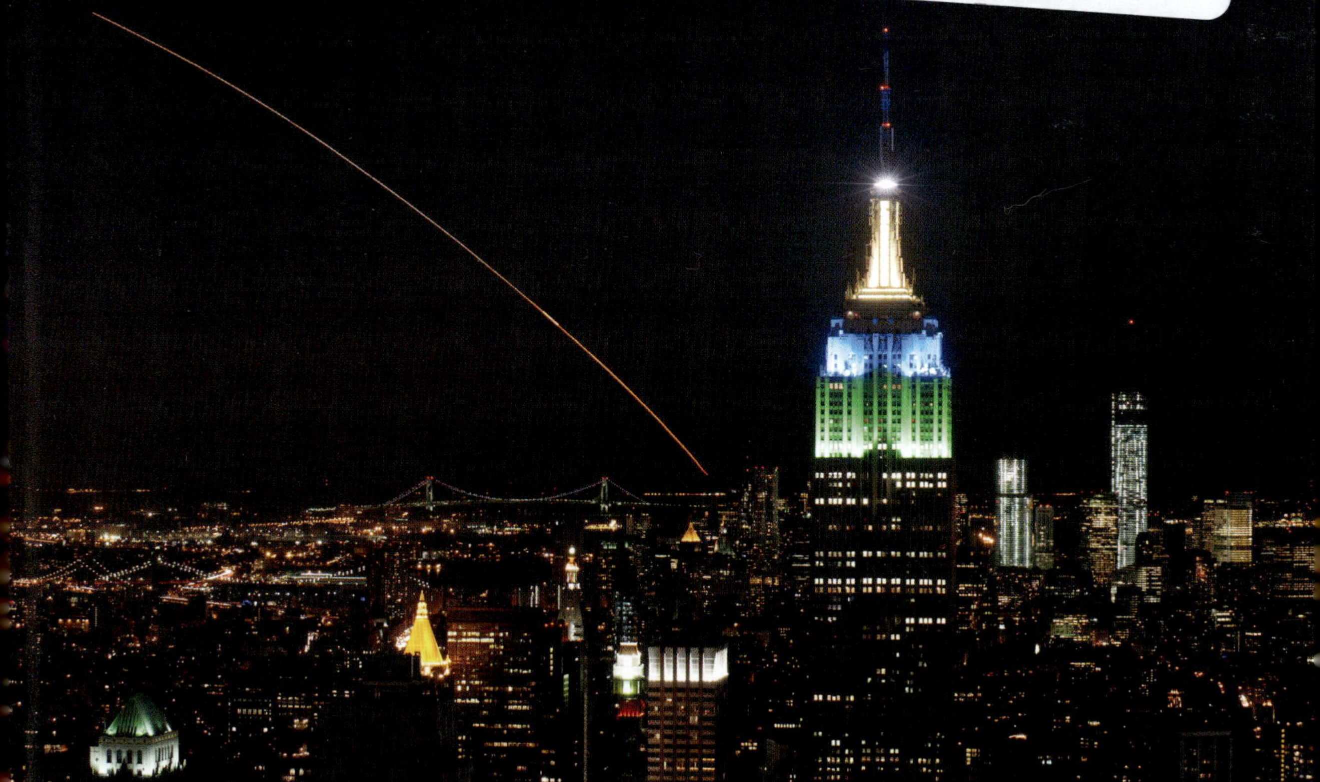

BEN COOPER

Amherst Media, Inc. ■ Buffalo, NY

(previous page) *In a view you would never expect for a space launch, a Minotaur V rocket takes off on the LADEE mission to the moon in 2013 from Wallops Island, Virginia, as seen in a time exposure from more than 200 miles north on the roof of Rockefeller Center in New York City.*

Published by:
Amherst Media, Inc.
PO BOX 538
Buffalo, NY 14213
www.AmherstMedia.com

Publisher: Craig Alesse
Associate Publisher: Katie Kiss
Senior Editor/Production Manager: Michelle Perkins
Editors: Barbara A. Lynch-Johnt, Beth Alesse
Acquisitions Editor: Harvey Goldstein
Editorial Assistance from: Carey A. Miller, Roy Bakos, Jen Sexton-Riley, Rebecca Rudell
Business Manager: Sarah Loder
Marketing Associate: Tonya Flickinger

ISBN-13: 978-1-68203-416-3
Library of Congress Control Number: 2019932345
Printed in the United States of America
10 9 8 7 6 5 4 3 2 1

CONTENTS

(below) Photos showing what the cameras, positioned at the launch pad and under rockets, go through after snapping the clear photos they get at first. Cameras are frequently pummeled with exhaust, water, sand, and mud.

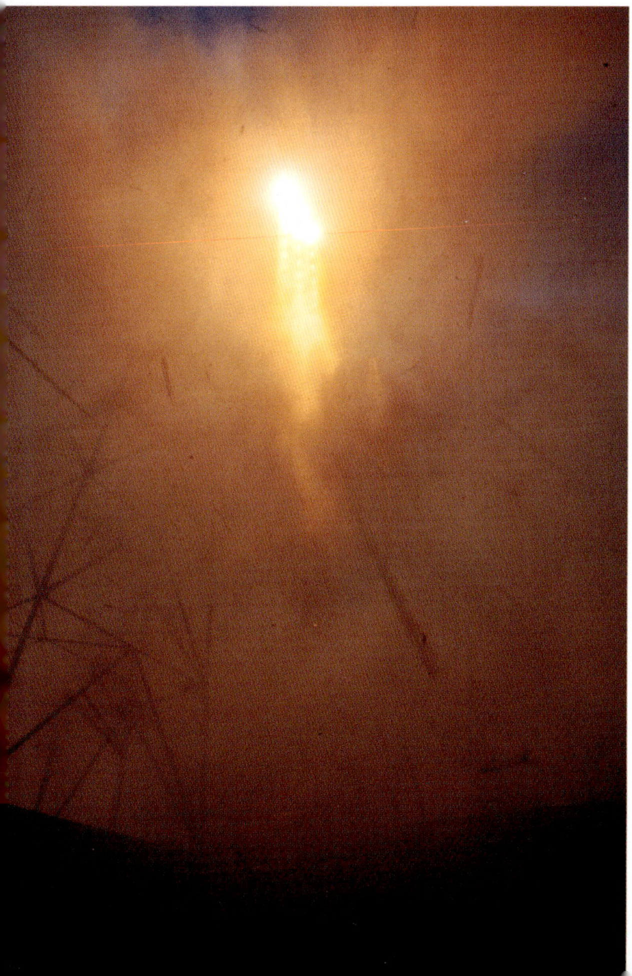

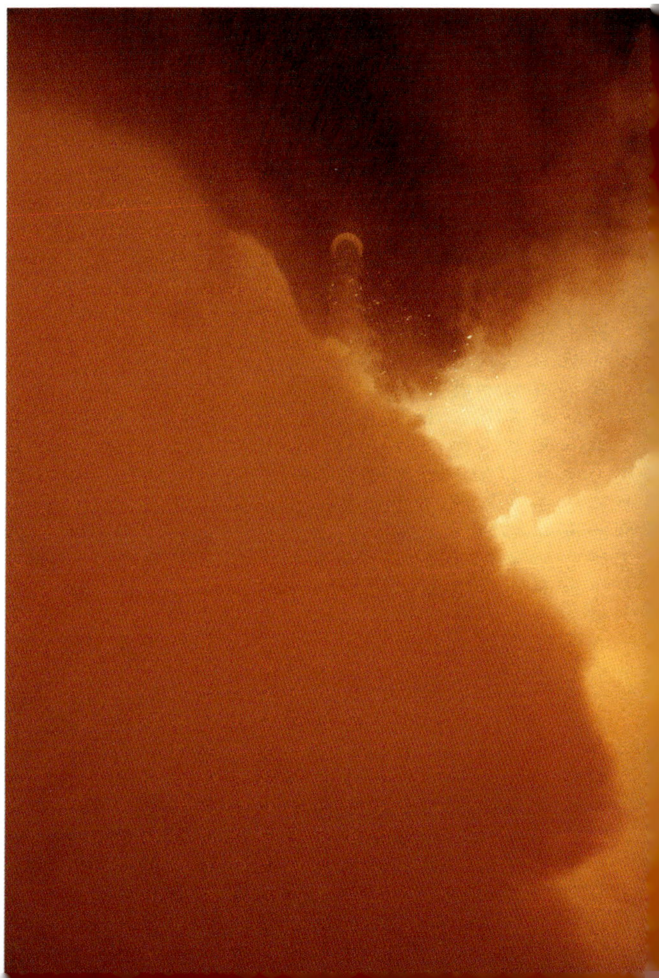

ABOUT THE AUTHOR

Ben Cooper is originally from Brooklyn, New York. He grew up a lover of both aviation and space, as well as photography, which he began to pursue around age eight. At fourteen, he traveled to Florida with his father to try to see a space shuttle launch for the first time. It took more than one try, but ever determined, when Ben finally saw one, he was hooked. From the very first attempt, Ben sought to get the perfect photo, going so far as to contact someone from NASA via the Internet for advice.

In 2003, looking to pursue a career in the space program, Ben attended Embry-Riddle Aeronautical University in Daytona Beach, Florida, for a degree in aerospace engineering, which he obtained in 2008. During his time here, just an hour north of Cape Canaveral, he continued to pursue hispassion of photographing launches. He soon had the privilege to photograph for prominent space news publications such as *SpaceflightNow, Aviation Week & Space Technology* magazine, *collectSPACE,* and others. By the time he was ready to graduate, he had a broad portfolio in photography and offers to apply for a NASA photographer position. He had the opportunity now to photograph the remaining space shuttle launches and their mission preparations for NASA directly, which he did from 2008–2011, when the aging vehicles were retired. Much of this work was critical to the lives of the astronauts on board, as NASA implemented broad documentation requirements for photography both on the ground and in space during missions following the Columbia accident in 2003. During the same period, new space launch companies were beginning to establish themselves and conduct their first launches in an era some have dubbed "New Space," and Ben was quickly asked to join in on documenting the first flights of this new era.

To learn more about Ben , see more of his work, and purchase prints, go to:

- www.LaunchPhotography.com/
- www.facebook.com/LaunchPhoto
- www.twitter.com/LaunchPhoto

Acknowledgments

This book is dedicated to all my friends and followers who have helped and supported me the last 20 years, not the least of whom is my wife, Katie, whom I met setting up cameras at a shuttle launch in 2010; my parents, Betsy and Marty and sister, Rachel; and great friends, especially photographers William G. Hartenstein and James Brown and writers Justin Ray and Robert Pearlman for giving me the opportunities to get started, and without whose mentoring, I might not be where I am today.

FOREWORD

It is a challenge to translate into words the tremendous explosive force released by a rocket launch. Journalists and authors, myself included, have tried to convey the sheer power of the liftoff to varying degrees of success, but no text quite compares to the sight of an impossibly bright, fiery flame thrusting a vehicle off Earth and into space.

And that is why what Ben Cooper does is so special.

Through his photography, we all get to experience the thrill of a launch, from vantage points where no human could ever safely stand and watch (let alone a camera survive; in the pursuit the photographs in this book, more than one camera and numerous lenses were destroyed).

Triggered by timers and the audio cues of the rockets' roaring blast, Ben's cameras have provided a front row seat to space history—to space shuttles taking astronauts to orbit, to new, reusable rockets blazing a path into space and to robotic planetary explorers leaving our world to probe the farthest reaches of our solar system.

And while there is a technological component to the images he captures, Ben is also an artist. His care in framing the rockets on their launch pads and finding ways to incorporate the often serene and fragile natural settings that surrounds these violent (if only for a few moments) locales, juxtaposes the height of human ingenuity with the core reason that we explore space—to improve our lives here on Earth.

I take a certain pride that I was able to help give Ben his start. But Ben's own "launch" was just the beginning. His work earned him a place among NASA's staff photographers and shooting for some of today's leading rocket companies to capture their official launch imagery.

So let the countdown begin! As you turn the pages of this book, appreciate the beauty, marvel at the power, and remember that behind every image is the artistry of launch photographer Ben Cooper.

—Robert Pearlman, space historian, journalist, and editor of collectSPACE.com

THE LATE SHUTTLE ERA
(1999–2011)

Prelaunch Milestones

Every shuttle mission began not with a launch, but with the landing of the mission before it.

The space shuttle was the world's first reusable space vehicle, flying 135 times between 1981 and its retirement in 2011. Five of them were built: Columbia, Challenger, Discovery, Atlantis, and later Endeavour, which replaced Challenger after its tragic loss in 1986. (Columbia would also be lost in 2003, leaving NASA with three remaining through 2011.) With four operating at a time, each would be inspected or refurbished between missions and launched again, usually a few months later. So a mission began immediately after landing on the runway at Kennedy Space Center, Florida, or Edwards Air Force Base, California.

The Shuttle Landing Facility

(below) Seen here, Atlantis is towed off the runway, known as the Shuttle Landing Facility (SLF), following its STS-129 mission in 2009.

From here, it is headed into one of the three Orbiter Processing Facilities (OPFs), the hangar where it will be serviced and checked for things like damage to any of its more than 26,000 heat-resistant silica tiles. This process would normally last a few months. The shuttle, properly known as the orbiter when not attached to its launching rockets, would then be rolled into the much larger Vehicle Assembly Building (VAB), attached to those boosters, rolled out to one of the twin launch pads at Cape Canaveral, and sent into space.

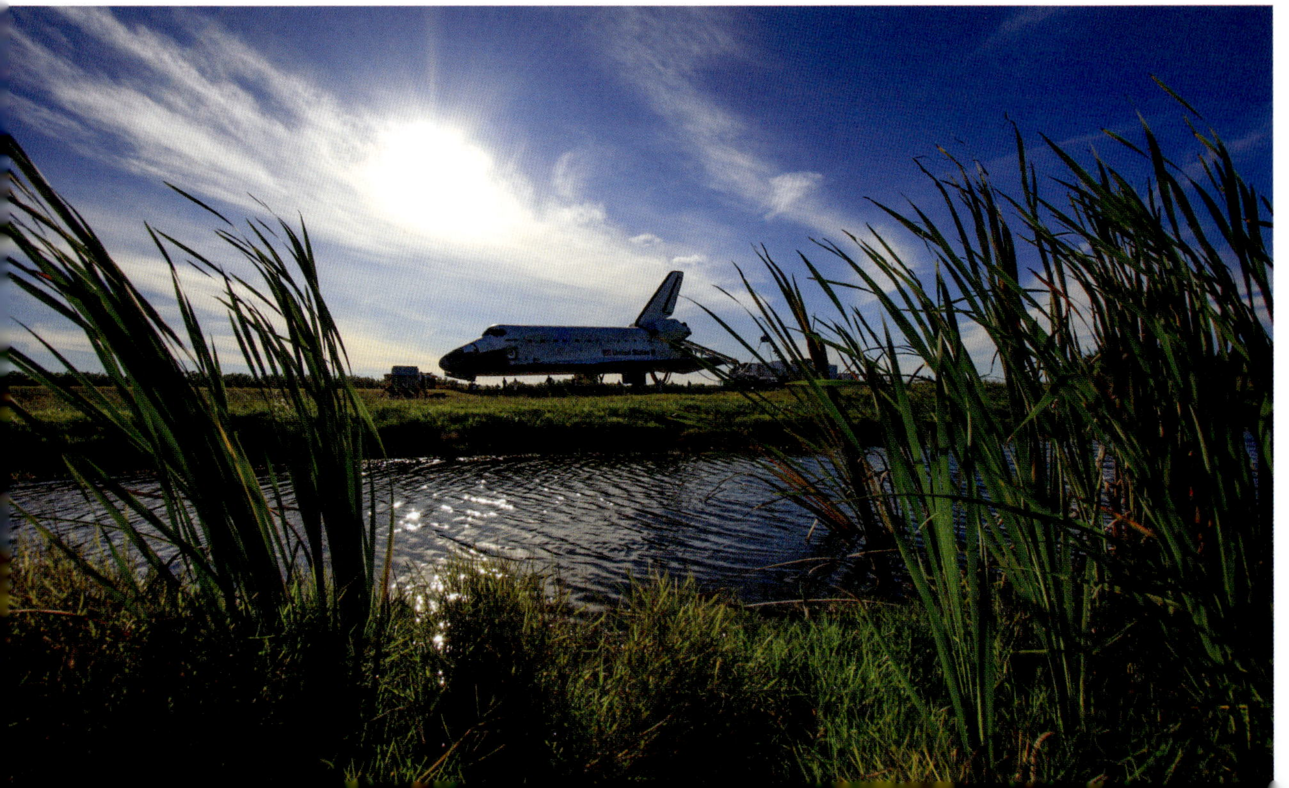

Inspection

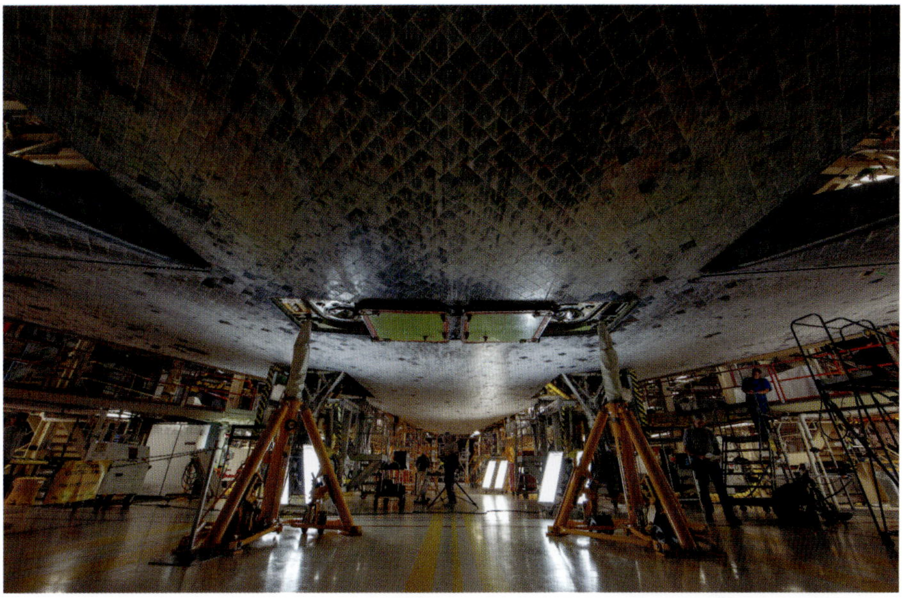

(top) Inside the OPF, technicians inspect every part of the complicated space vehicle. The outside of the orbiter is covered in thermal protection to safeguard it against the extreme heat of reentry upon return to Earth, and the extreme hot-and-cold temperatures of space, whether facing toward or away from the sun. The difference can be up to 400 degrees Fahrenheit, while reentry can exceed 3,000 Fahrenheit. The majority of the protection are thousands upon thousands of small silica tiles, black and white in color.

Seen here, the bottom of the orbiter is covered in all black. Elsewhere, white areas could be tiles, blankets, or other special materials developed for the program. The streak-like markings in this photo are scorch marks from the high-speed reentry into Earth's atmosphere.

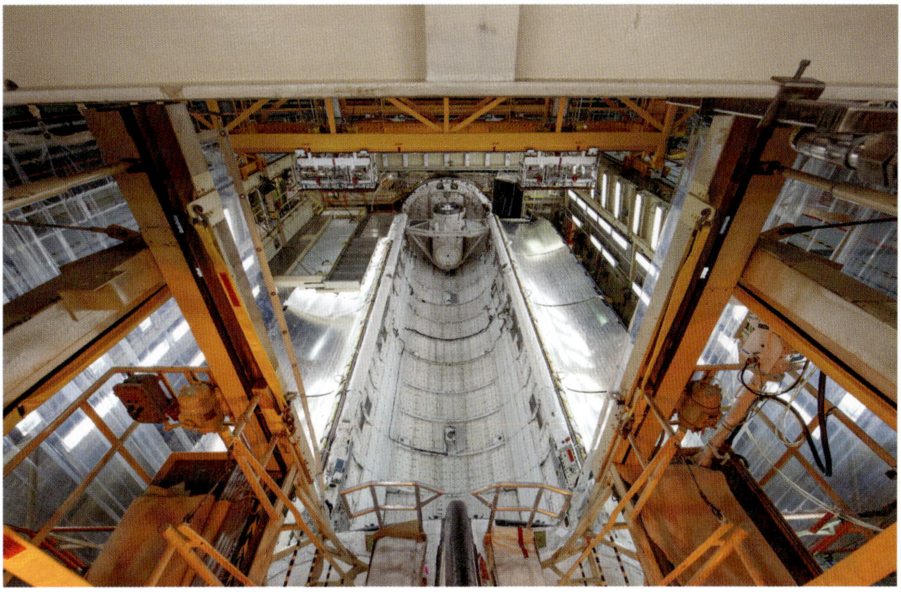

A View from Above

(bottom) Looking down from the roof of the building, the empty cargo bay of Discovery is visible, along with the open bay doors (shiny aluminum on either side), as well as the vertical stabilizer (tail), barely visible at the bottom of the image.

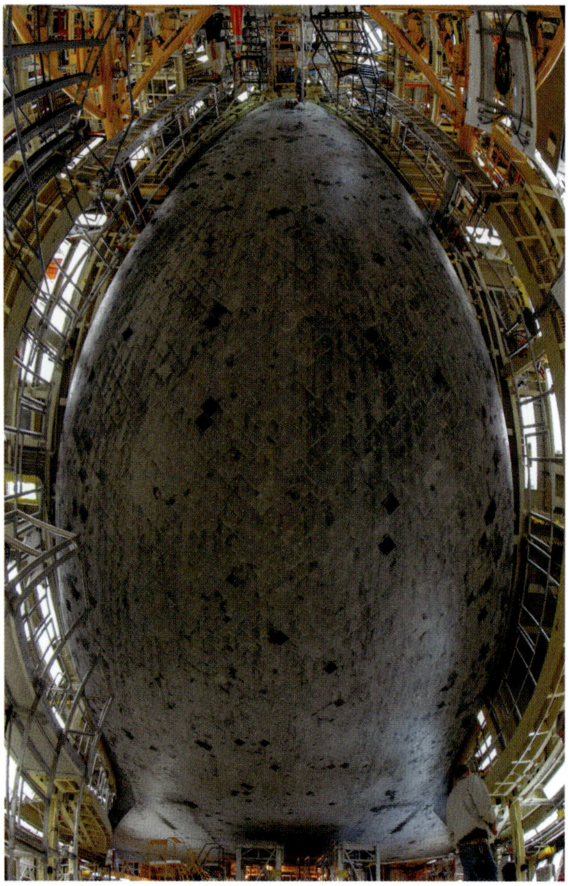

An Unusual View

(above) An unusual fish-eye view shows the entire belly of Discovery and the majority of the nearly 26,000 heat tiles that cover the airframe. Between missions, every tile was not only inspected, but also photographed, something implemented in the latter third of the program following the Columbia accident. Photos would be taken section by section in a grid-like pattern, and later would be used to compare to similar photos taken once the shuttle was launched into space. This would tell the astronauts and mission controllers whether any of the fragile tiles were damaged during the trip between the ground and space.

Discovery is surrounded on all sides by scaffolding. So well was an orbiter hidden by floors and platforms designed to fit its shape that when you walked into an OPF, someone who did not work here would not know a space shuttle was present unless told.

Vehicle Assembly Building

(following page) While the orbiter is being processed in the OPF, next door in the huge VAB, the booster rockets that will take the 100-ton spaceship into orbit around the Earth are being assembled.

The VAB was completed in 1966 to support the Apollo program and the huge Saturn V rocket that took men to the moon. It was the world's largest building by volume for years (it is still in the top seven as of 2018), and continues to hold the record for largest one-story building on earth, standing 525 feet tall.

It is so cavernous that the enormous, 15-story orange fuel tank that the shuttle used looks small when hanging from the crane inside. Very slowly, it is lifted out of a checkout cell on one side of the building and carried over to the other, where it is attached first to the two white solid-fuel boosters on its sides.

This photo was taken from about halfway up the building in height.

The VAB was completed in 1966 to support the Apollo program and the huge Saturn V rocket that took men to the moon.

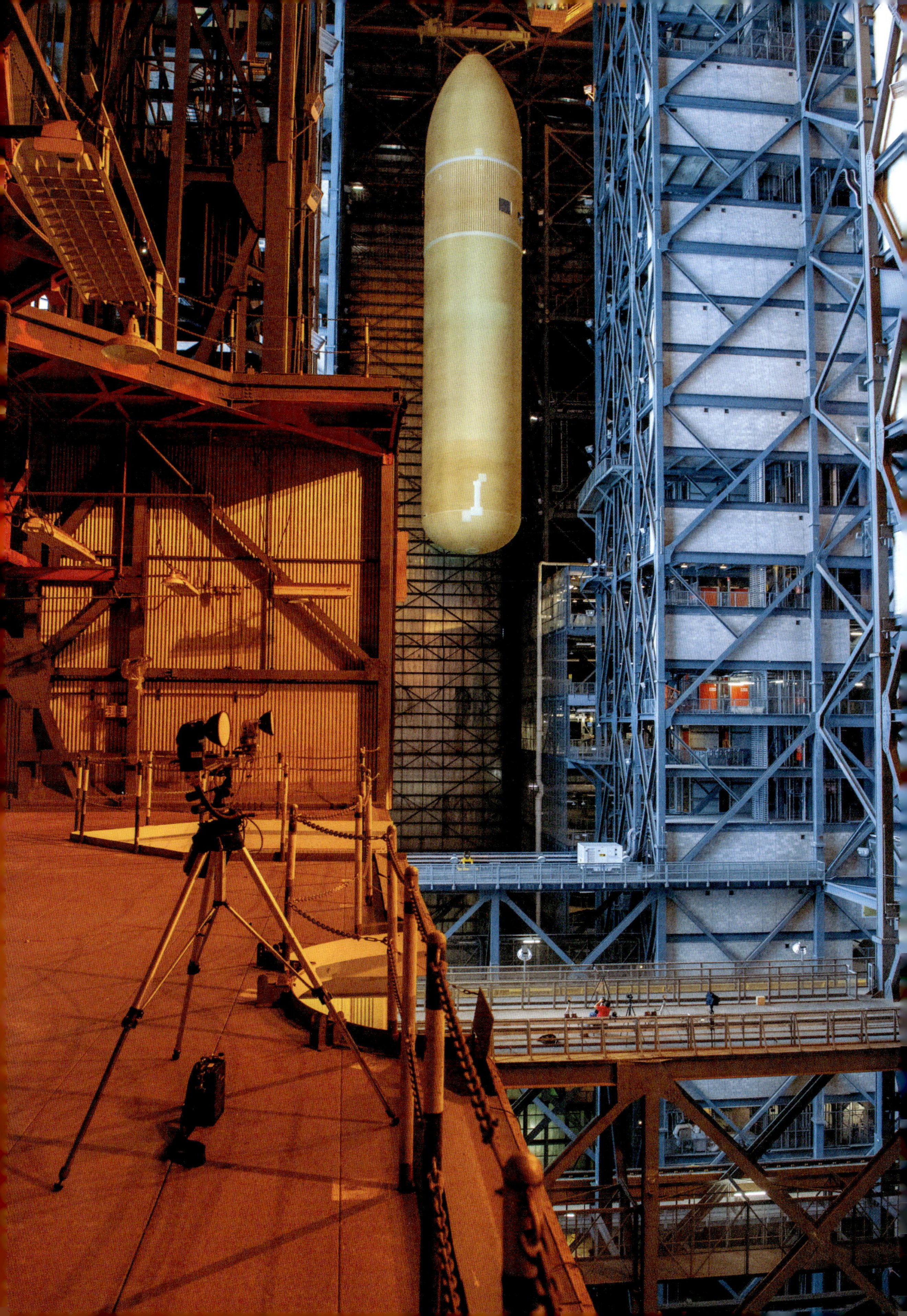

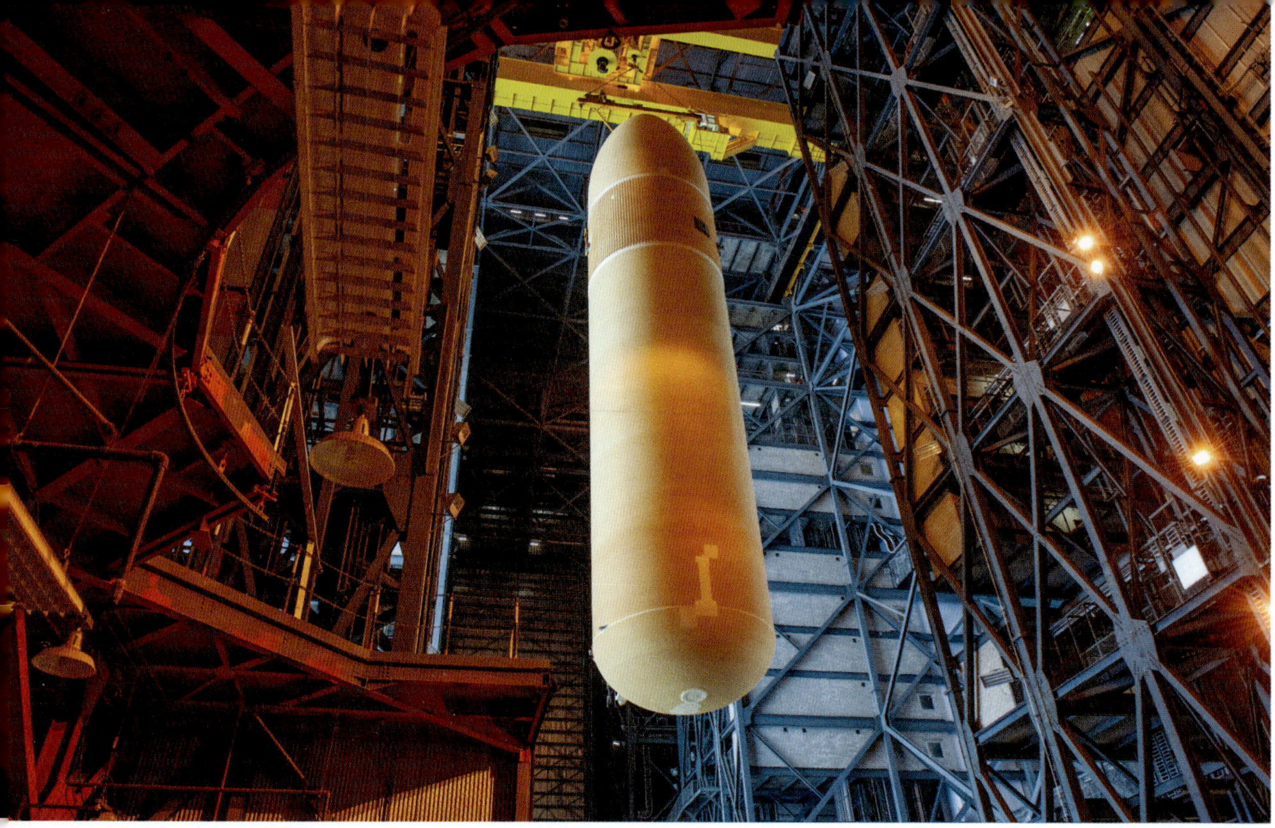

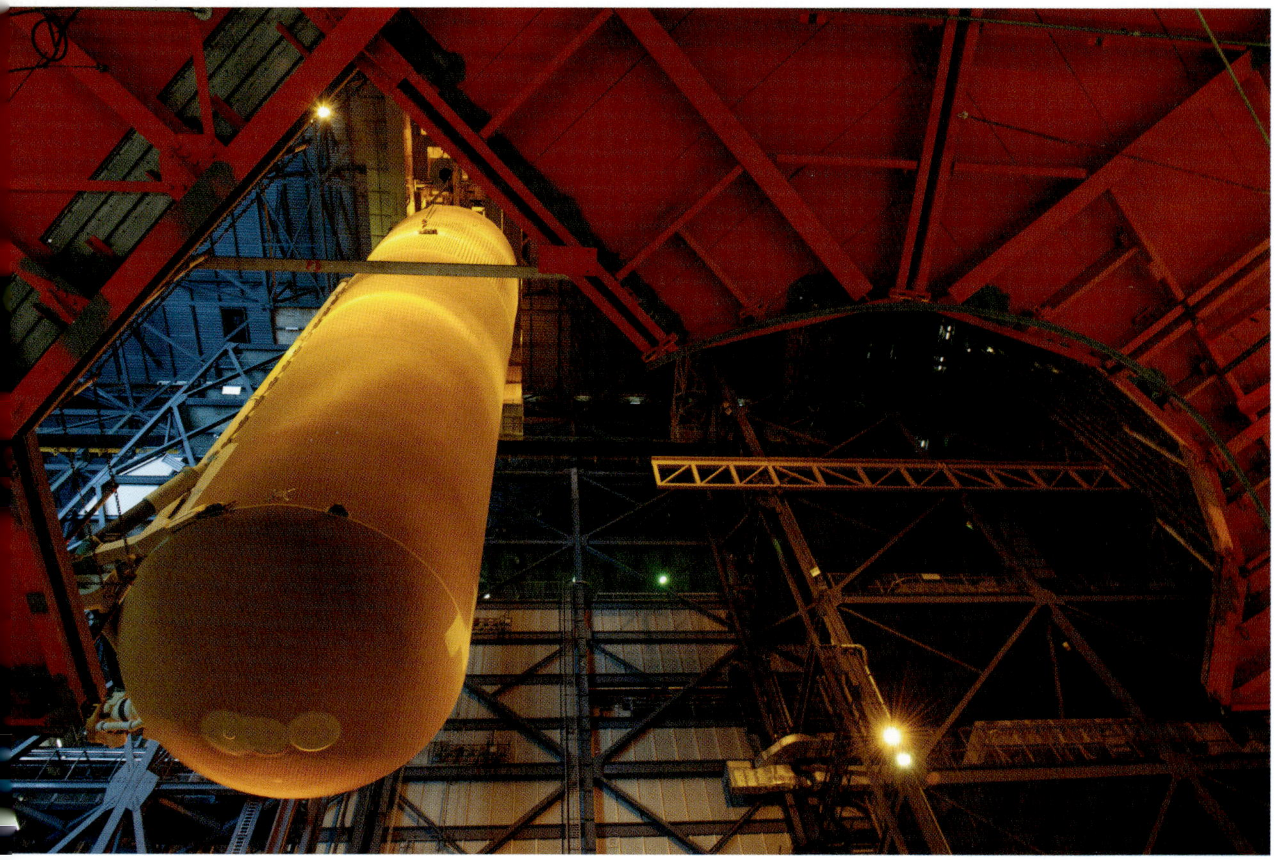

Inside the VAB

(previous page, top) Multicolored beams and floors made of steel exist throughout the fortress that is the VAB. Many areas of the building are still shaped for components of the Saturn V moon rocket, and painted red, as was the style of the launch facilities in the 1960s.

Lowering the Apollo

(previous page, bottom) Unusually shaped red flooring left over from the Apollo program frames the orange External Tank (ET) as it is lowered into position for an upcoming shuttle mission.

External Fuel Tank

(right) The external fuel tank, which will hold enormous quantities of liquid hydrogen and liquid oxygen during launch, is carefully attached to the Solid Rocket Boosters (SRBs) that flank either side of the stack. The rocket assembly is now ready to have the orbiter itself attached to it. Kennedy Space Center workers provide a nice size perspective.

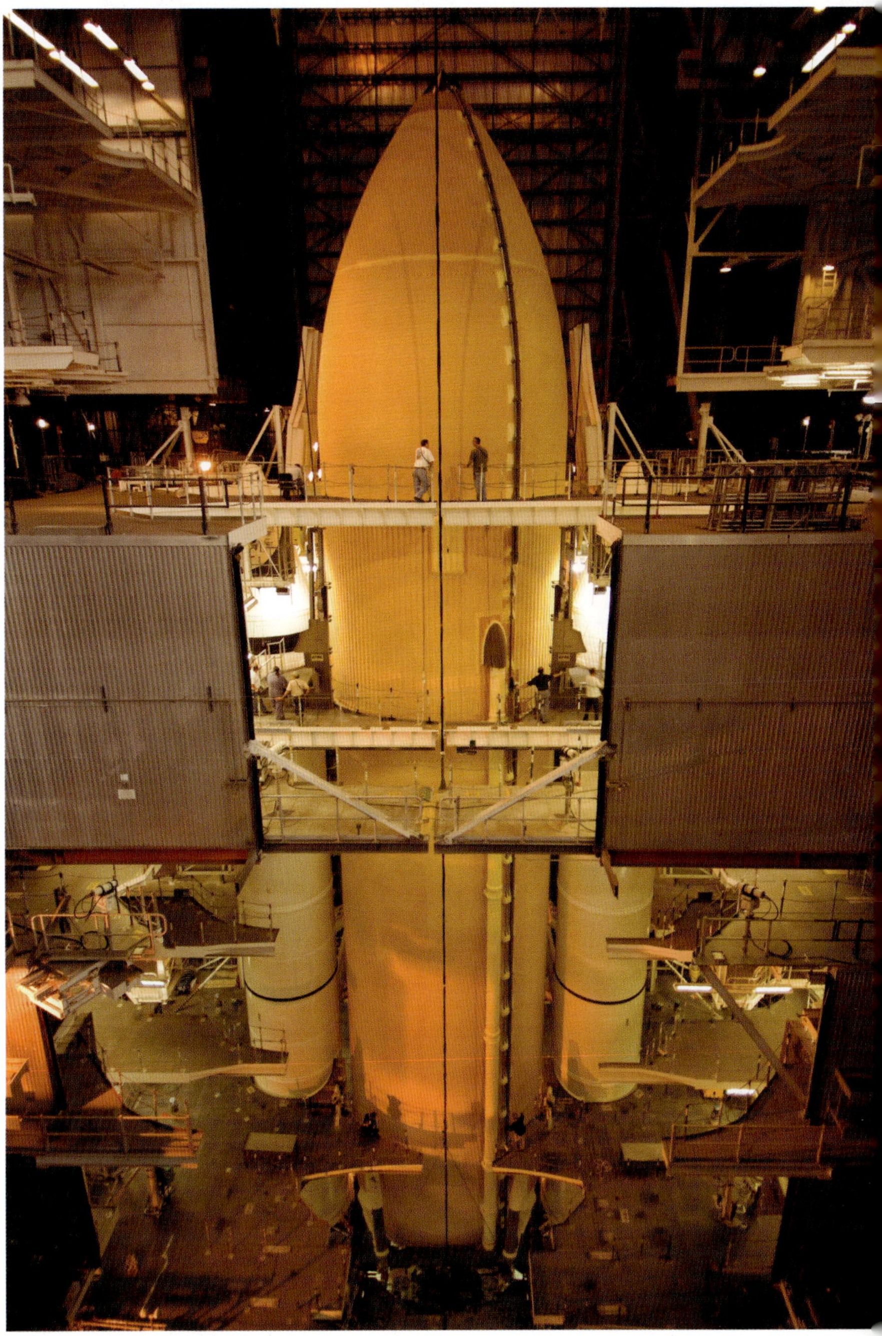

The rocket assembly is now ready to have the orbiter itself attached to it.

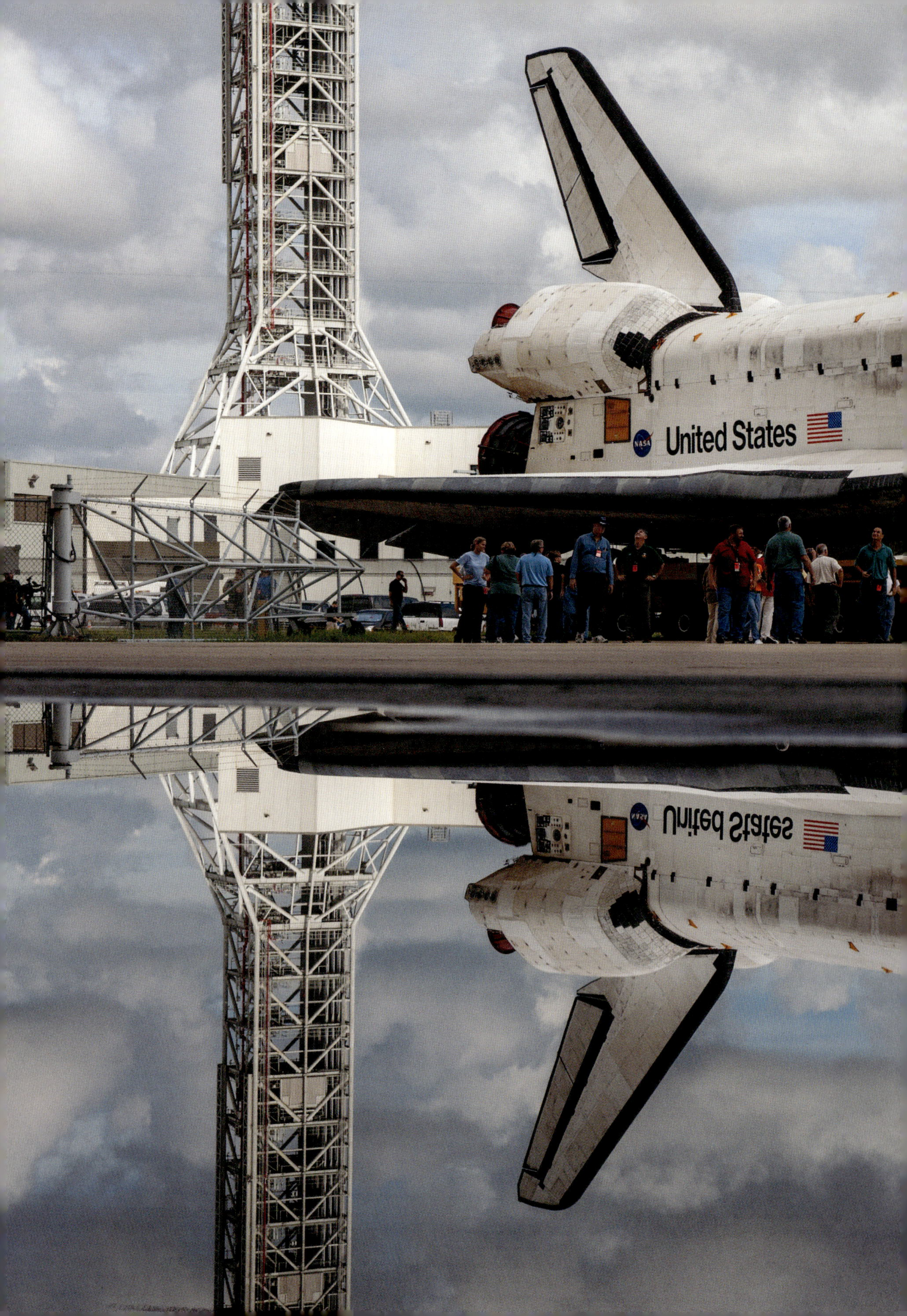
United States
NASA

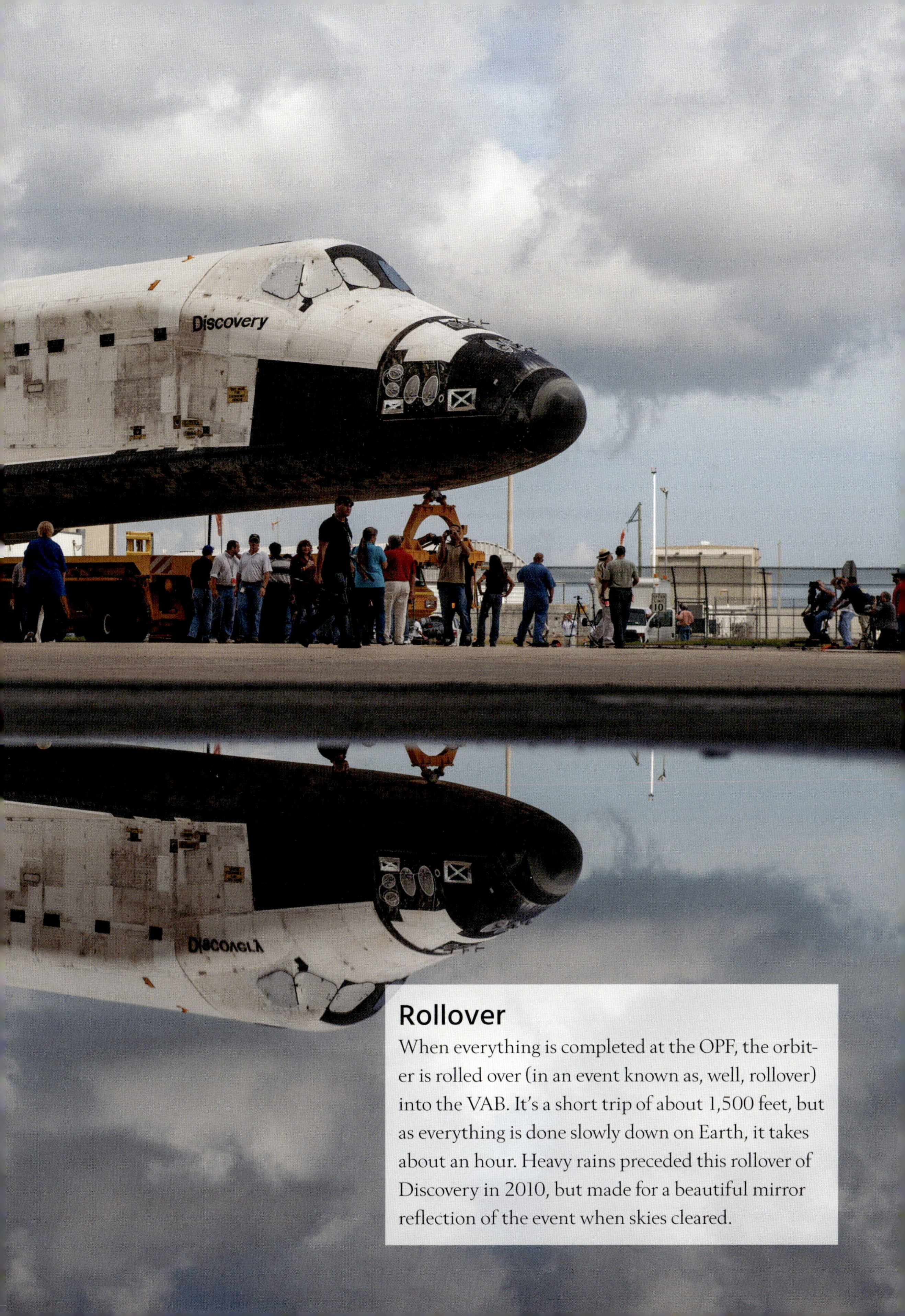

Rollover

When everything is completed at the OPF, the orbiter is rolled over (in an event known as, well, rollover) into the VAB. It's a short trip of about 1,500 feet, but as everything is done slowly down on Earth, it takes about an hour. Heavy rains preceded this rollover of Discovery in 2010, but made for a beautiful mirror reflection of the event when skies cleared.

Atlantis

(top) An aerial view of the rollover of Atlantis for the final space shuttle mission, STS-135, in 2011. This photograph was taken from the roof of the 525-foot VAB.

(bottom) Another view of Atlantis, this one captured inside the VAB.

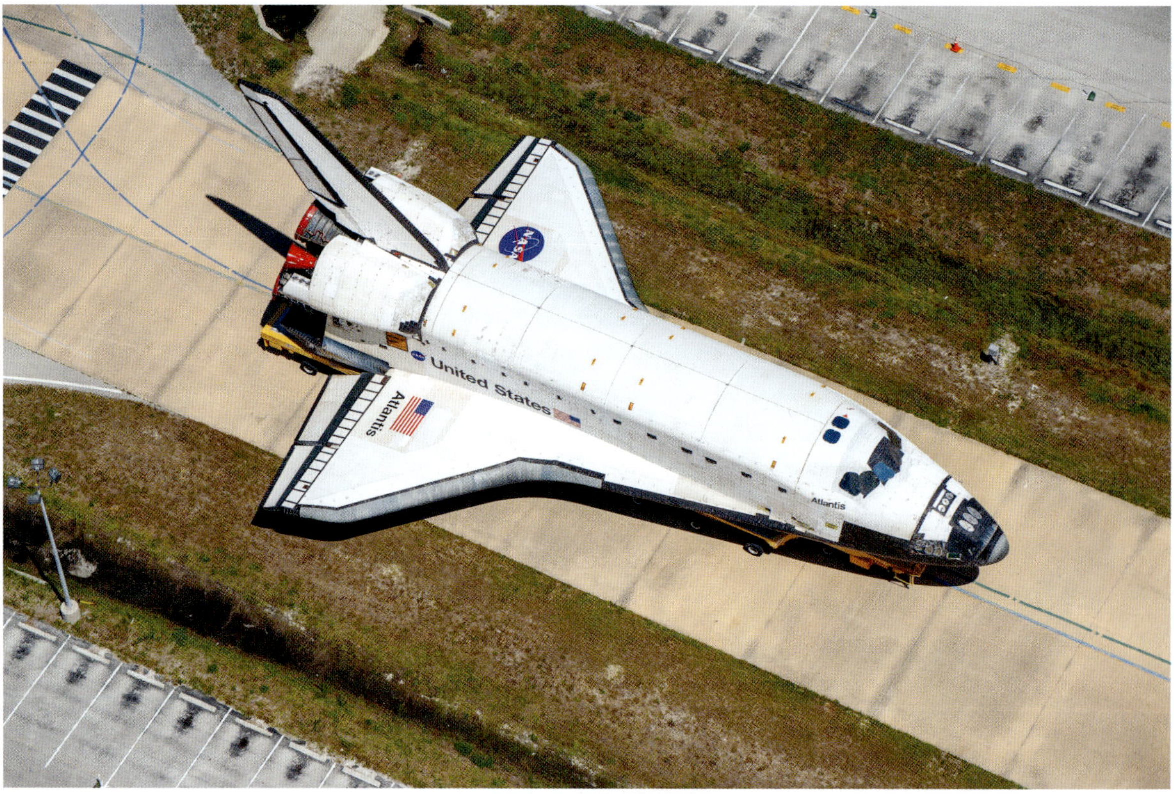

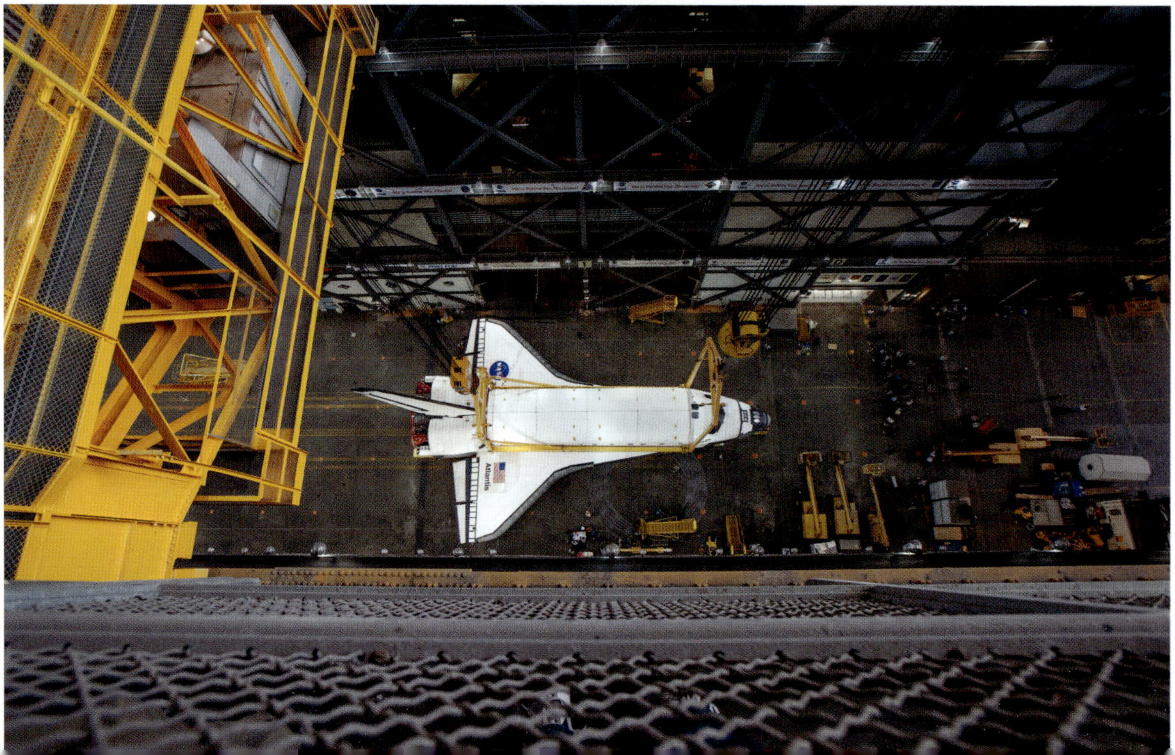

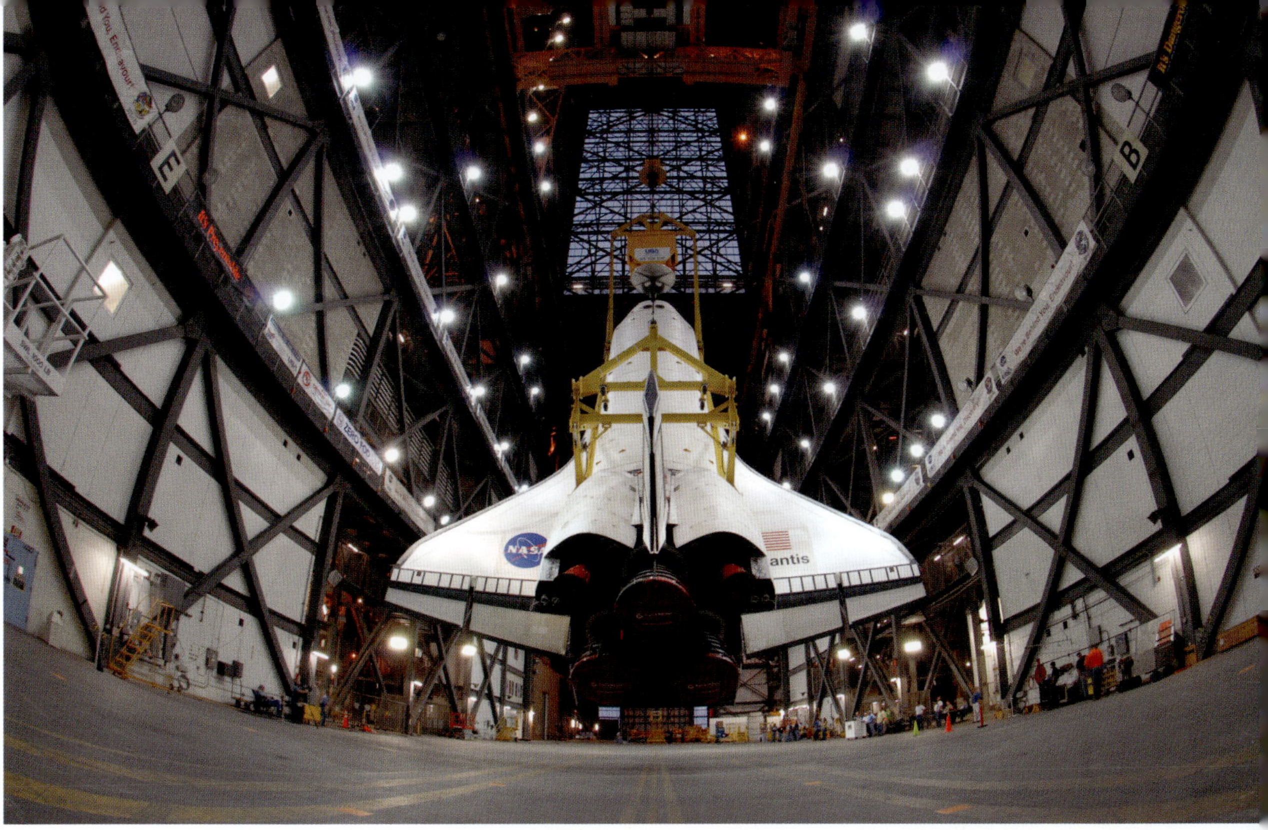

Slow and Steady

(top) Once the orbiter reaches the VAB, a crane capable of lifting as much as 400 tons of weight (the orbiter itself weighs about 100 tons) is attached, and a delicate ballet begins as the nose is brought up and the tail down, such that the vehicle is in the vertical position for attachment to the boosters. The entire process can take 24 hours, with movements of less than a foot per minute at times.

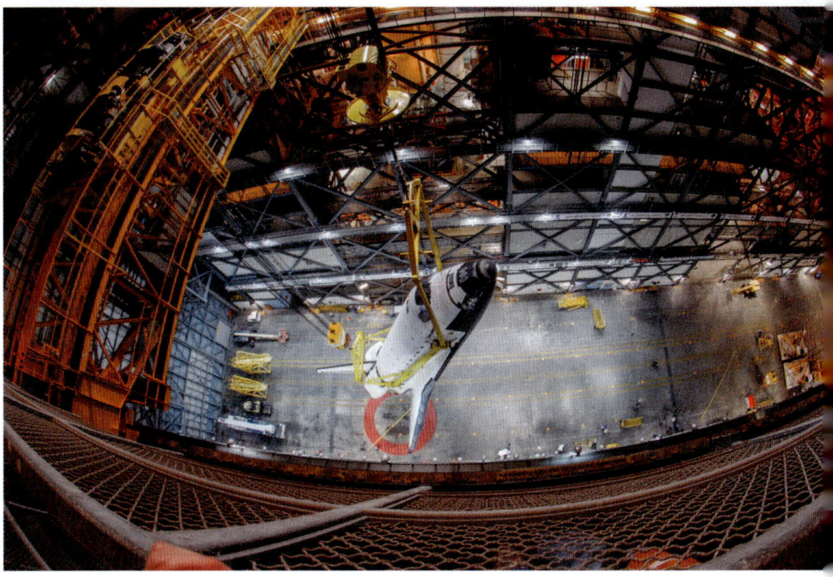

Unique Perspective

(bottom) A unique overhead angle again taken with a fish-eye lens provides a perspective seldom seen.

Once the orbiter reaches the VAB, a crane capable of lifting as much as 400 tons of weight is attached.

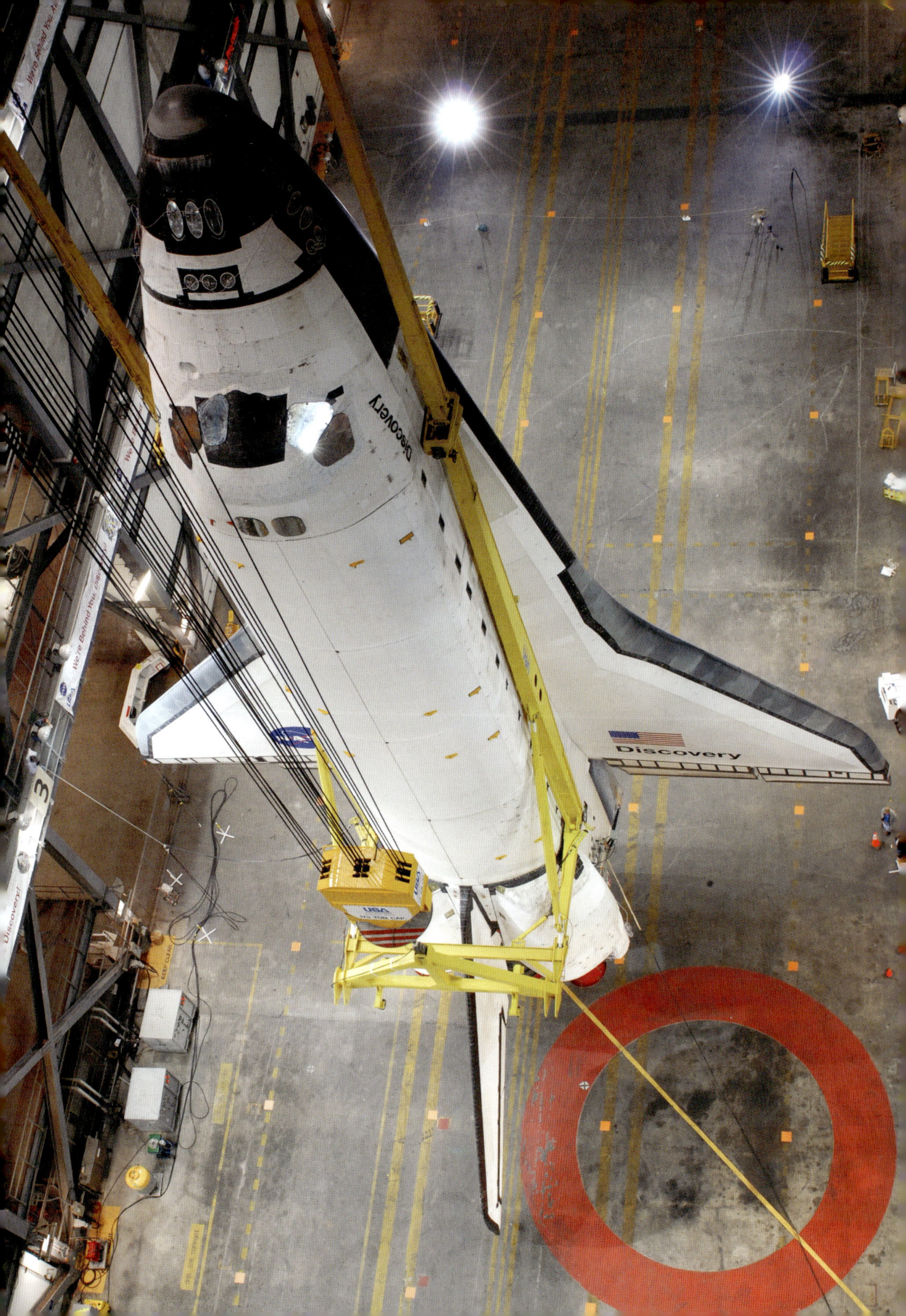

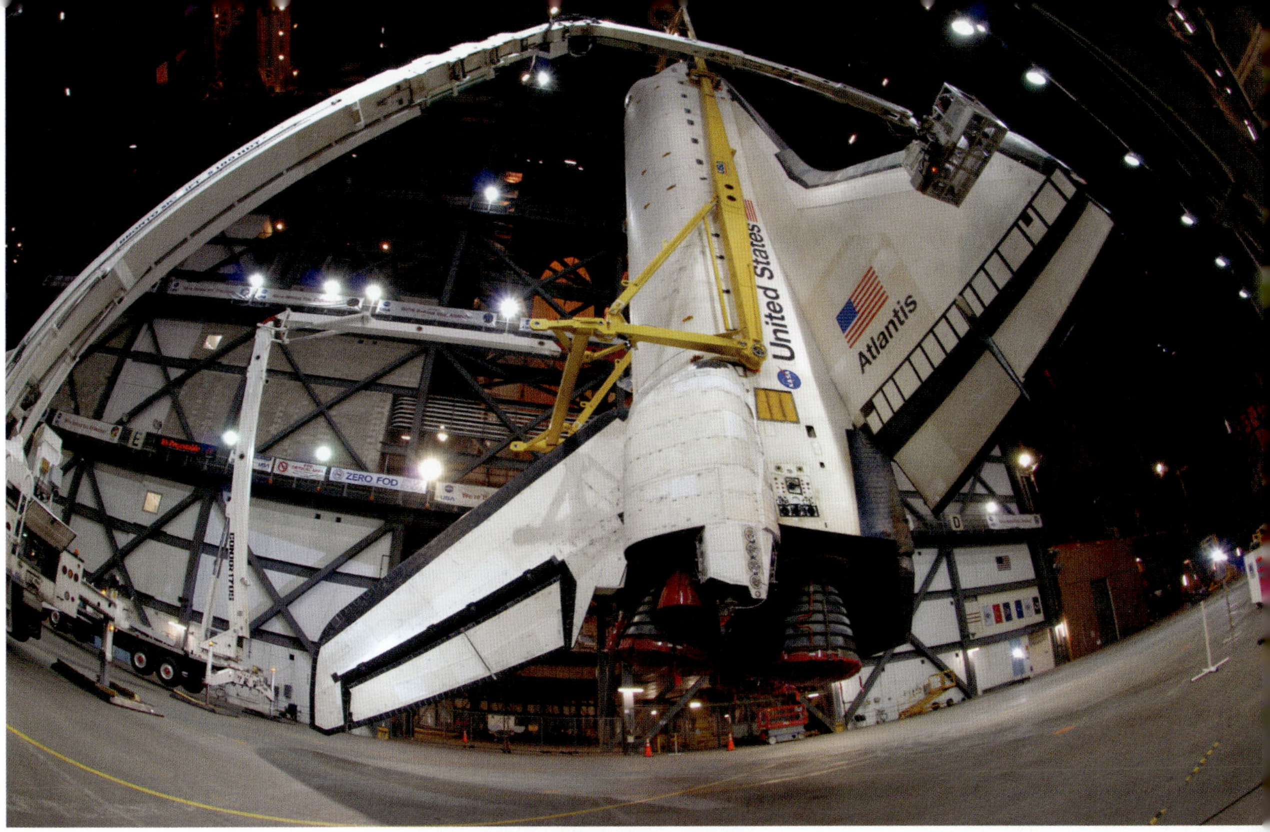

A Slow Climb

(previous page) Discovery is vertical and hanging from the VAB crane as it begins the slow climb toward the roof. The design of the VAB is such that there is a center, open area called the transfer aisle, and four open high bays on each side of the building (two on the east side, two on the west). To get between the transfer aisle and high bays, the orbiter and any other component must be lifted up and over a dividing wall and then lowered back down for attachment.

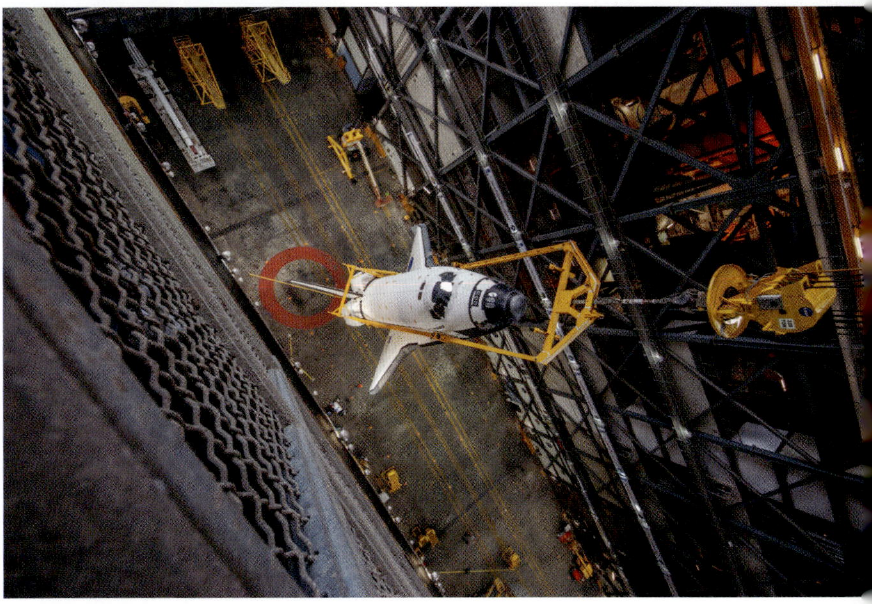

Unique Photo Opportunities

(top and bottom) A fish-eye view of the orbiter Atlantis after going vertical in the VAB. On the cherry pickers, photographers shoot angles of some of the wing's leading-edge panels that are not accessible when horizontal.

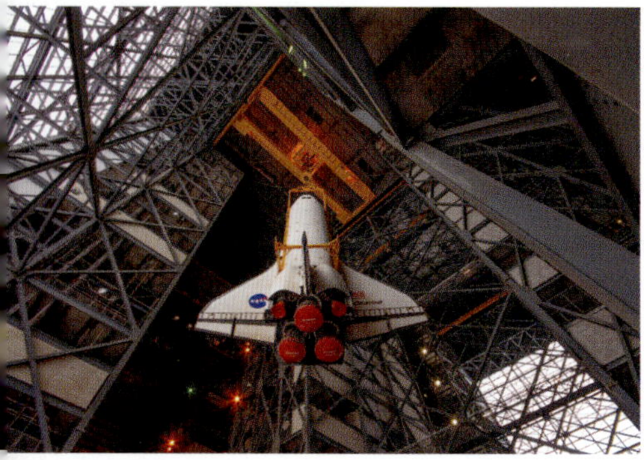

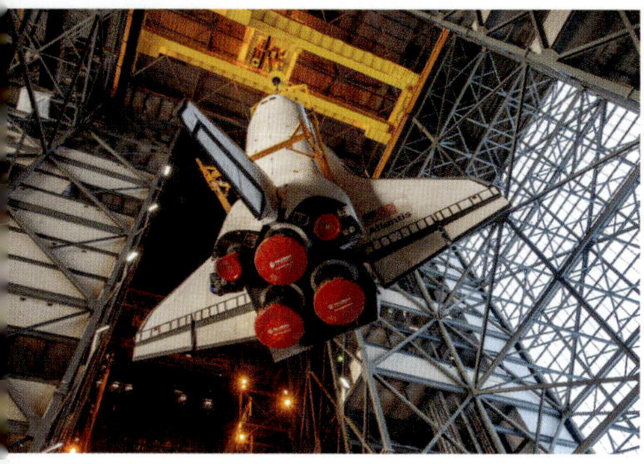

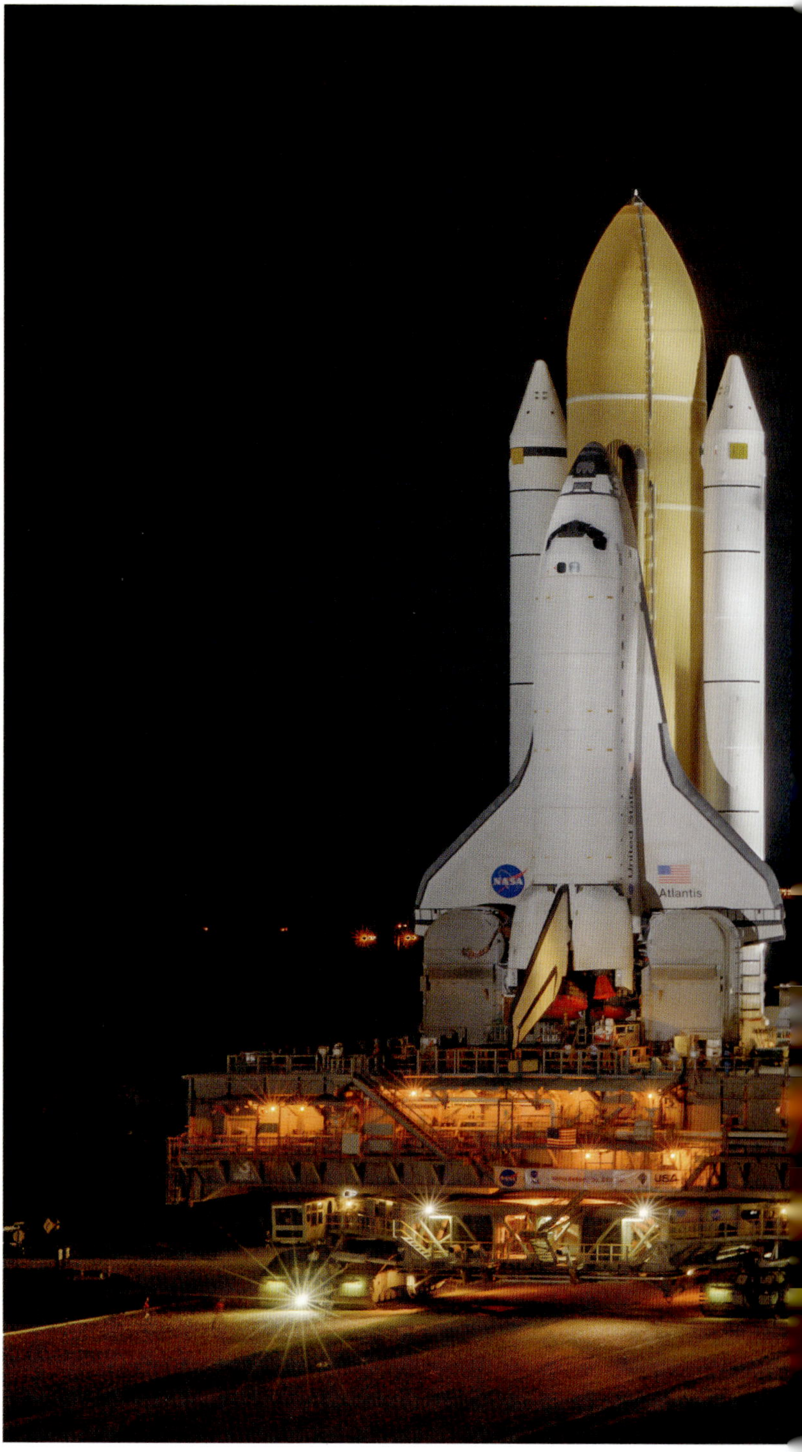

A Careful Ascent

(top and bottom) The orbiter is now slowly lifted to the roof of the VAB, turned on an angle so its wingtips fit through the opening to the bay, and lowered back down.

A Big Event

(right) A space shuttle rollout could be a big event. For the final rollout of the program seen here, thousands of spectators, mainly employees and their families, plus other guests and VIPs, lined the outside of the VAB to wish Atlantis a safe trip.

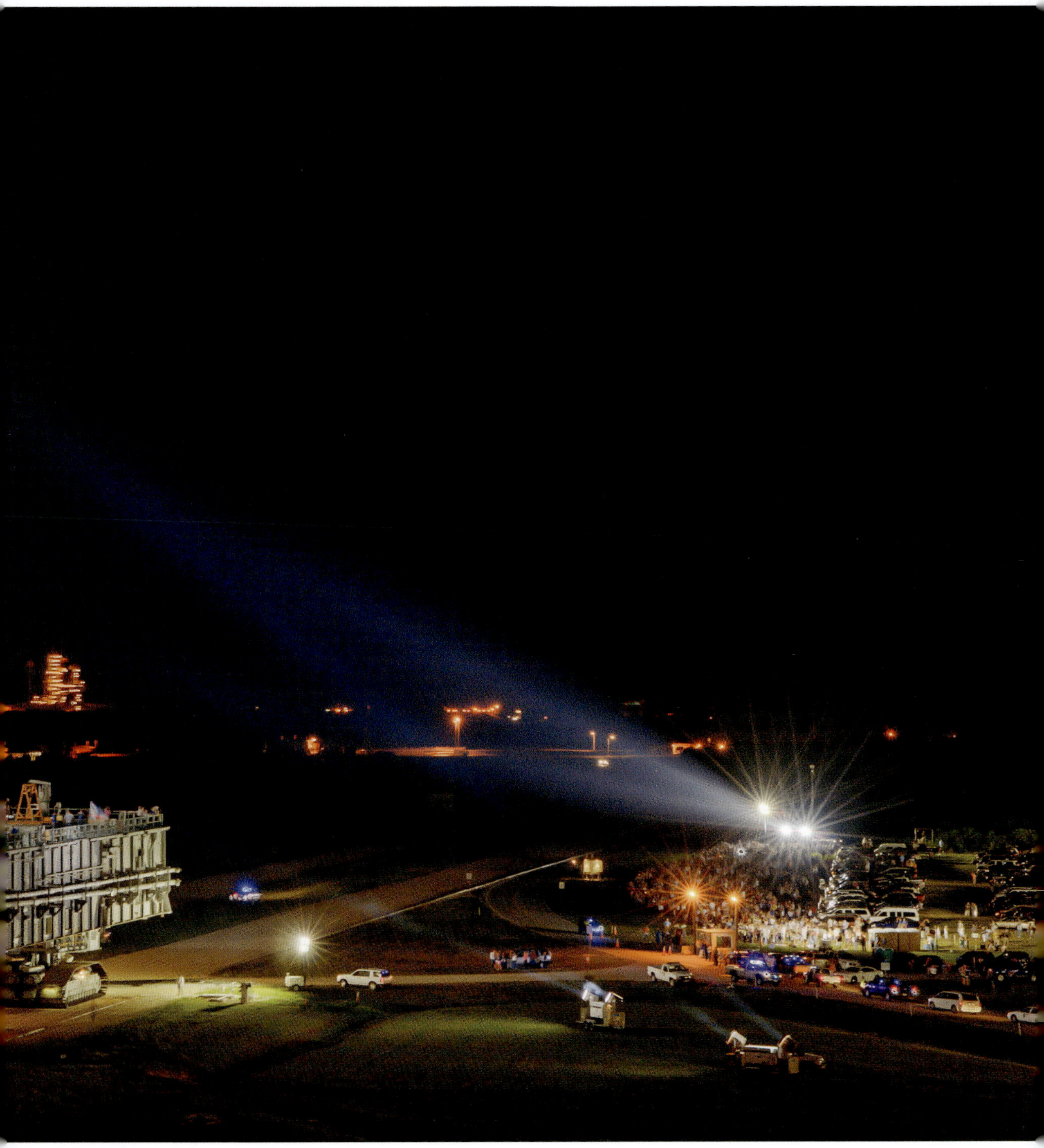

Ready to Roll

(following page spread) The shuttle stack is complete; it can now be called the space shuttle. In this rarely seen view, captured from the VAB's rafters, the space shuttle Endeavour is complete and ready to be rolled out to the launch pad.

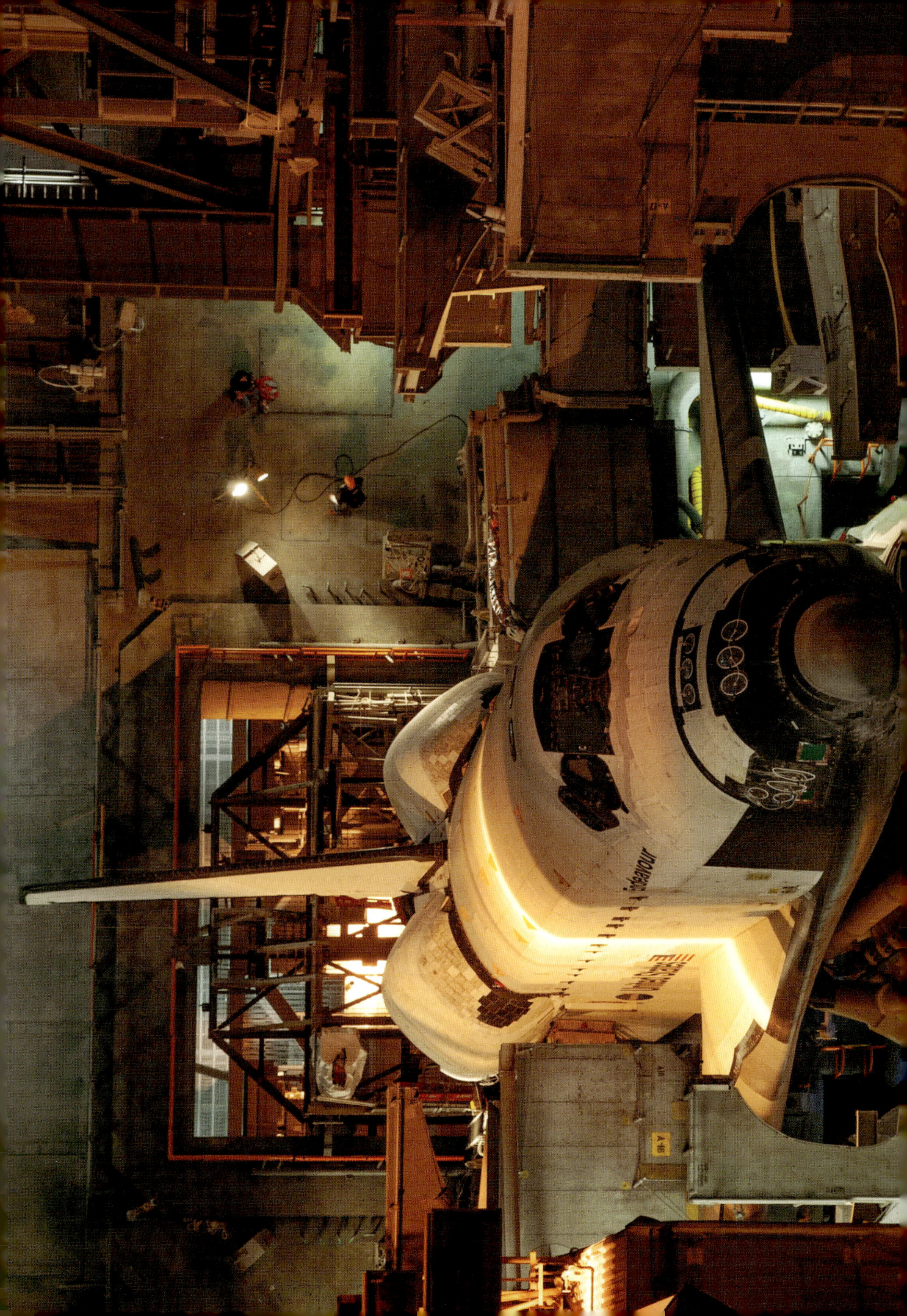
Endeavour

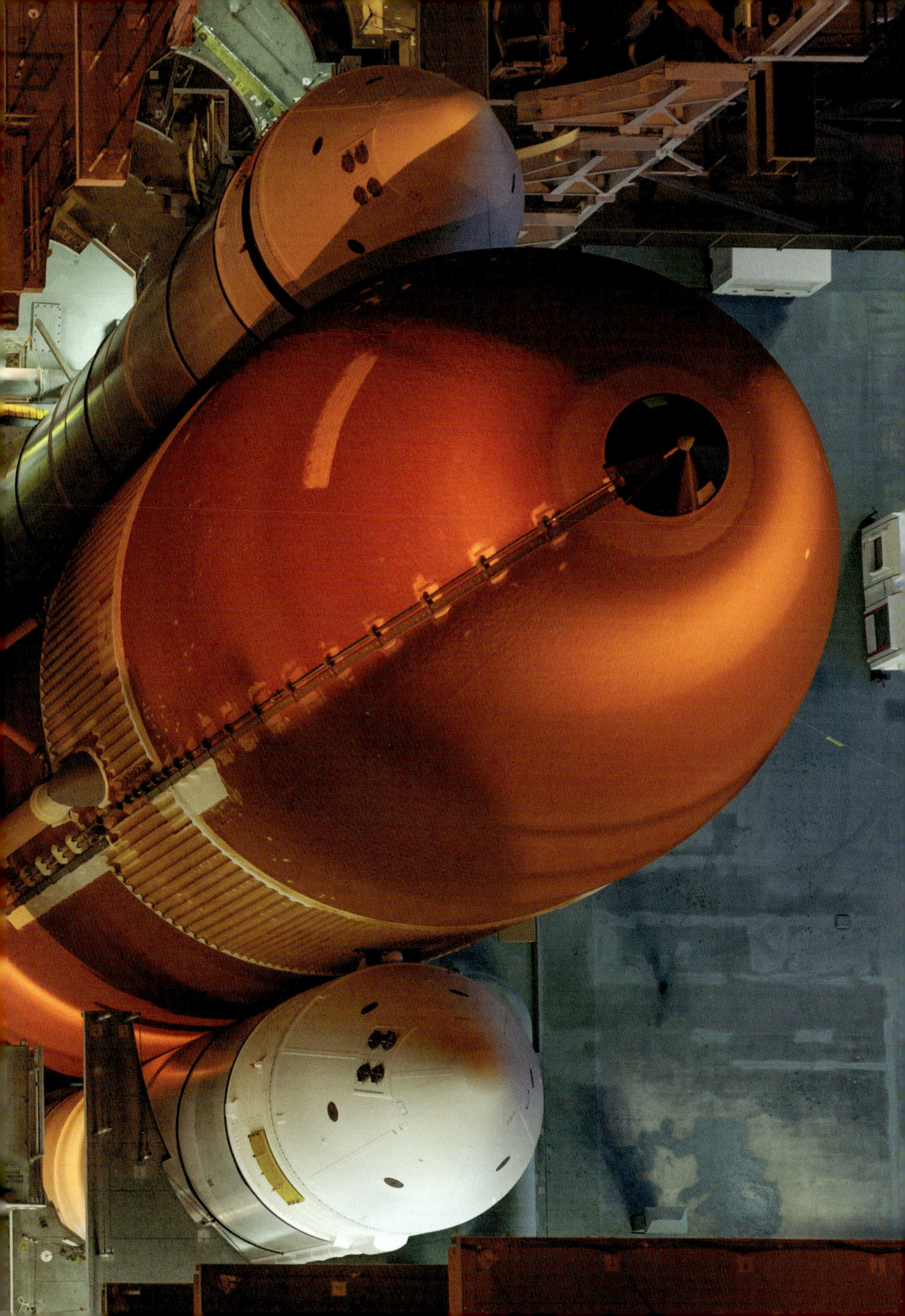

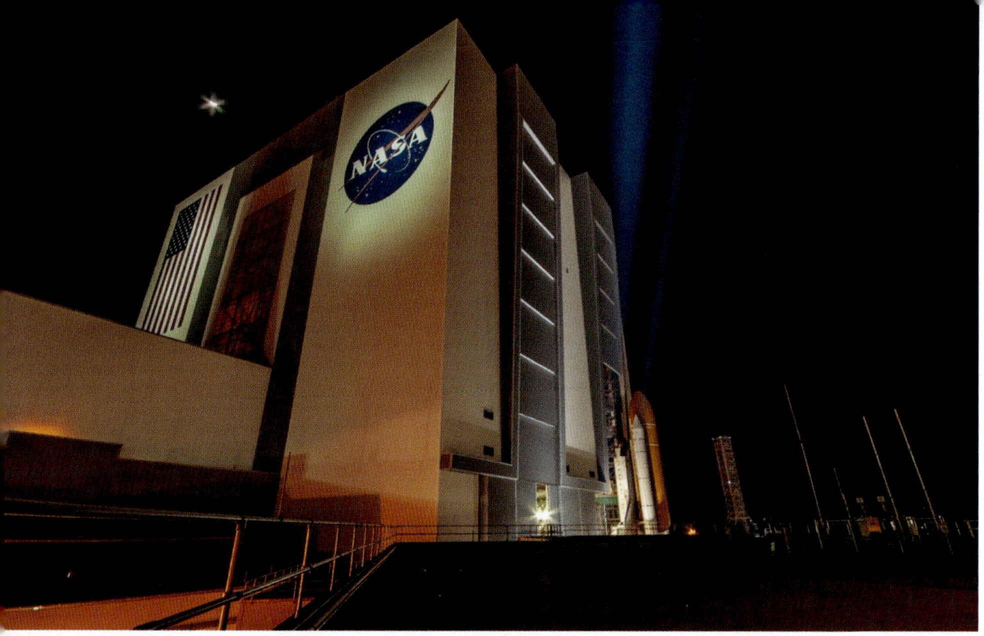

Emergence

(top) The space shuttle Endeavour emerges from the cavernous VAB as it begins the three-to-four-mile trek to the launch pad. Despite its size, standing more than 22 stories tall itself, the shuttle stack appears dwarfed by the 52-story-tall VAB.

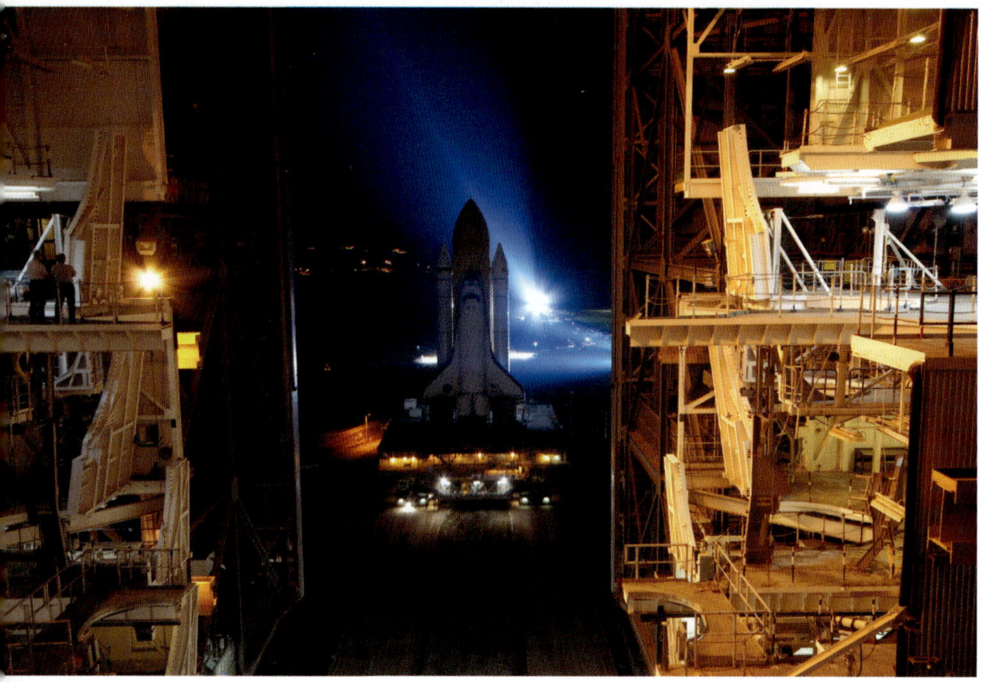

Pride

(bottom) Technicians look on as Atlantis departs the VAB, silhouetted by beams from xenon spotlights set up for the event.

A Lengthy Trek

(following page, top) Rollouts of the space shuttle typically began at midnight or before first light. The trip to launch pad 39A or 39B would take at least six to eight hours, at a top speed of just one mile per hour.

Heavyweights

(following page, bottom) The shuttle Discovery nears launch pad 39B at dusk, the end of a long day that began a little later than usual. The space shuttle, 184 feet tall, is mounted to the Mobile Launch Platform (MLP), which then sits atop the crawler transporter, one of the world's largest motor vehicles. The crawler and MLP together weigh more than 15 million pounds, even without the stacked space shuttle.

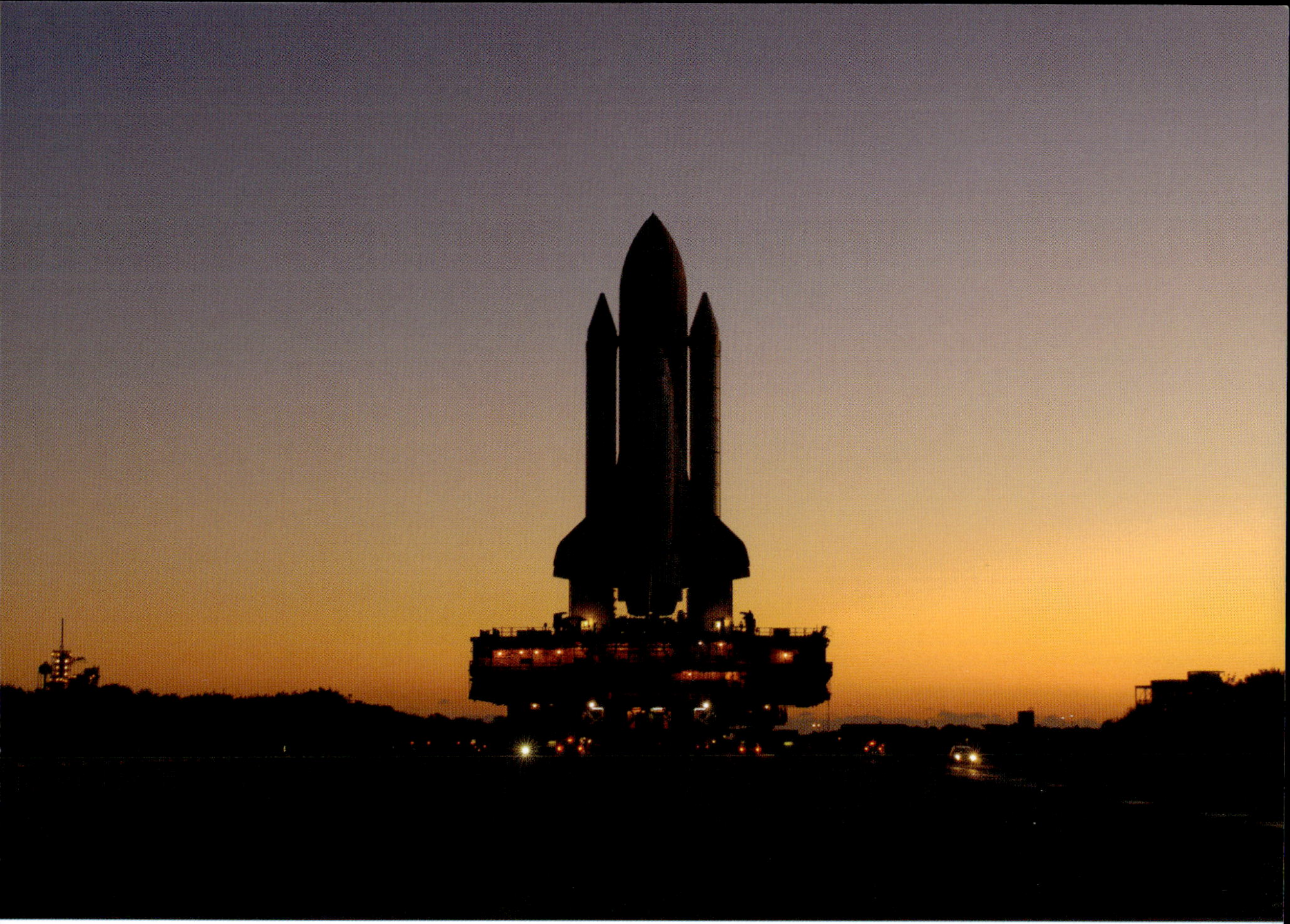

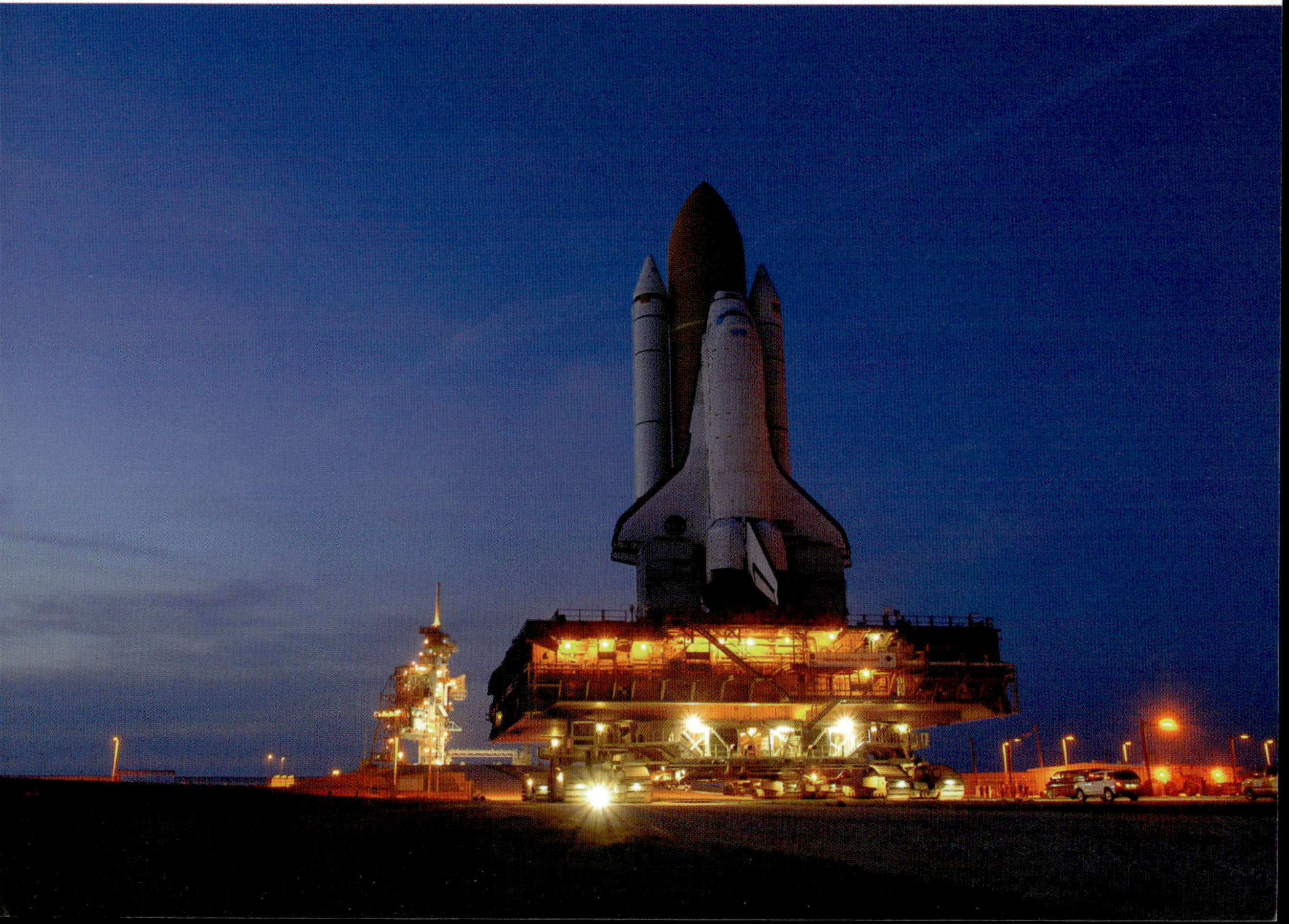

Around the Bend

Framed by bars on a nearby viewing platform, space shuttle Discovery curves around a bend in the crawlerway, the point at which one way heads to pad 39A while the other, a left turn, heads to 39B. Discovery was preparing to return the shuttle fleet to flight in 2005 following the Columbia accident more than two-and-a-half years earlier.

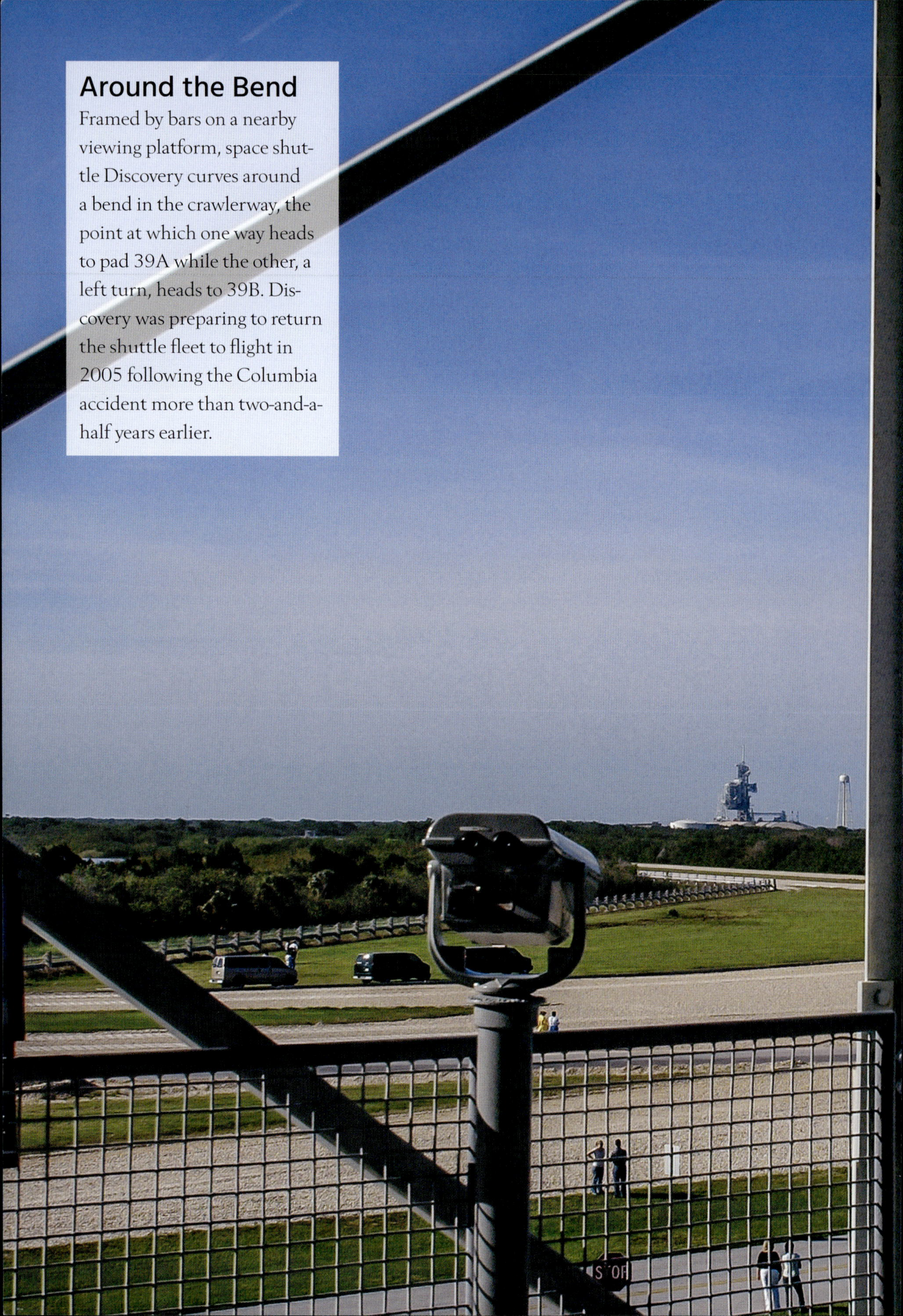

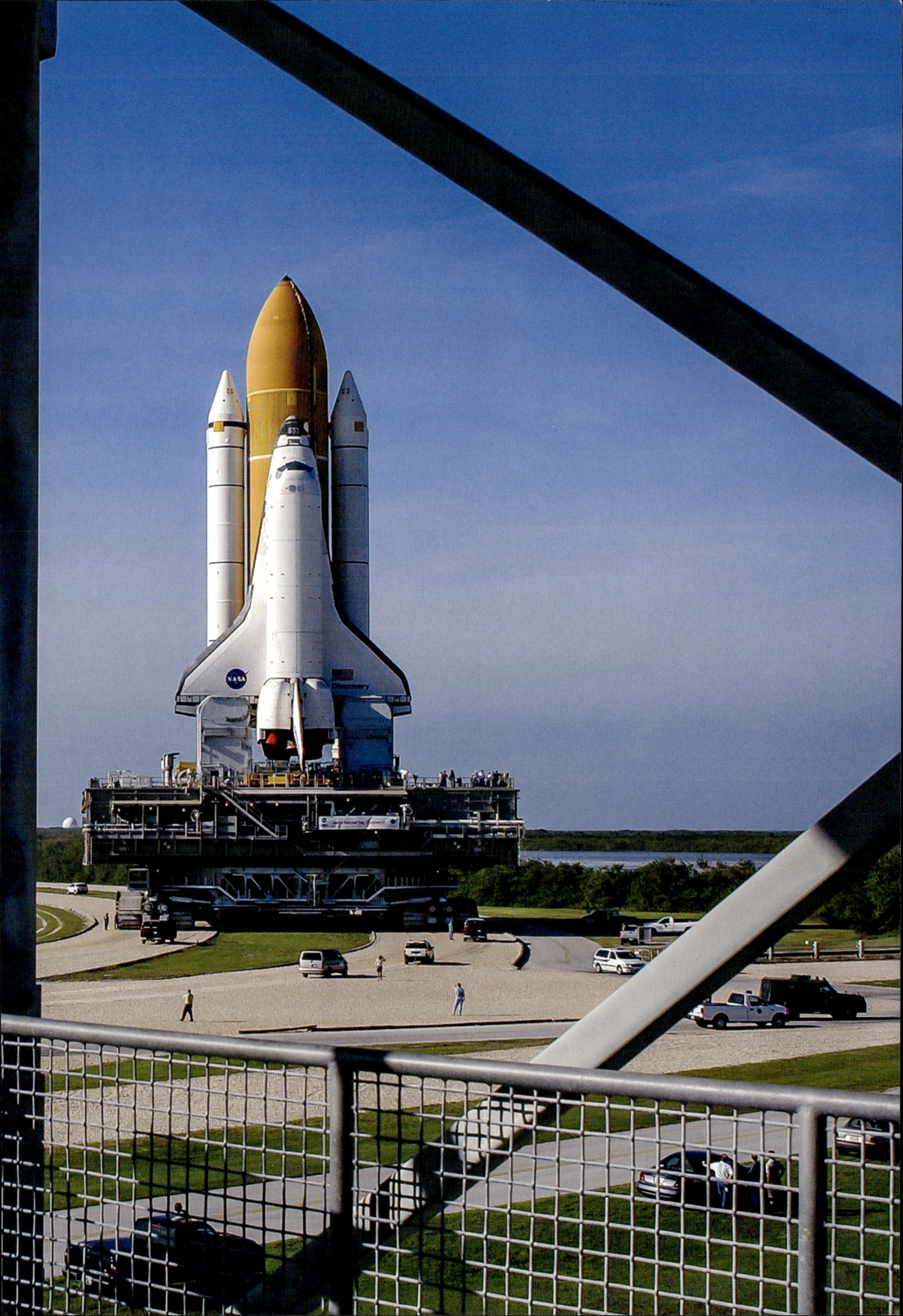
NASA
Discovery

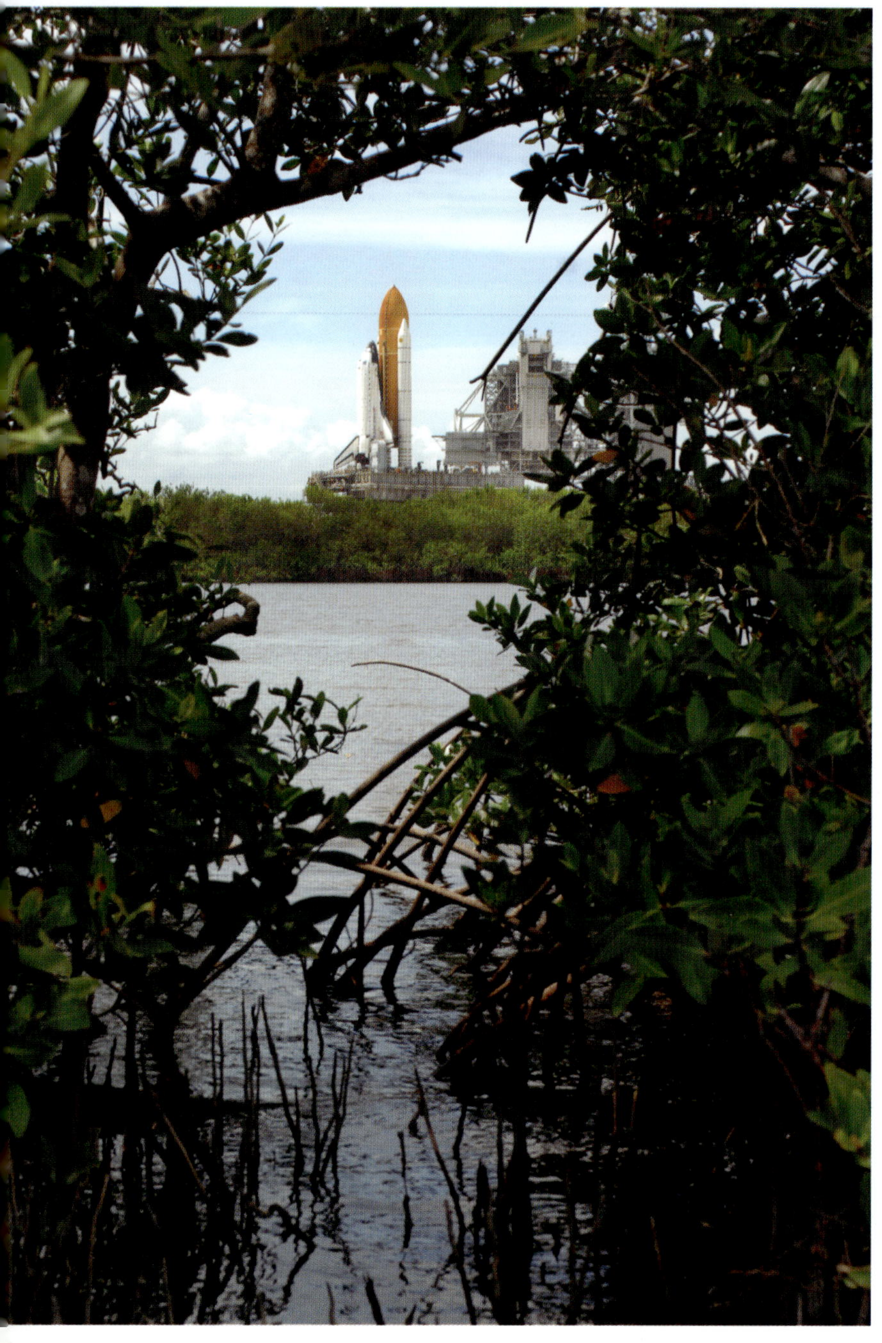

Merritt Island

(left) The Kennedy Space Center is completely surrounded by Merritt Island National Wildlife Refuge, created as a buffer zone for the hazardous operations that are rocket launches. Mangrove swamps make up a large part of the refuge, and the brackish-water vegetation with its showy roots can frame photos beautifully.

A Rocky Road

(following page) The 18-million-pound shuttle stack rides down the crawlerway on its way between the VAB and launch pad. Consisting of seven feet of rock and stone piled atop each other, the topmost layer is composed of loose river rocks, which take the brunt of the weight. Each of the individual tank-like treads on the crawler weighs one ton, and the force of the stack rolling over the rocks can pulverize them into dust.

The Kennedy Space Center is completely surrounded by Merritt Island National Wildlife Refuge . . .

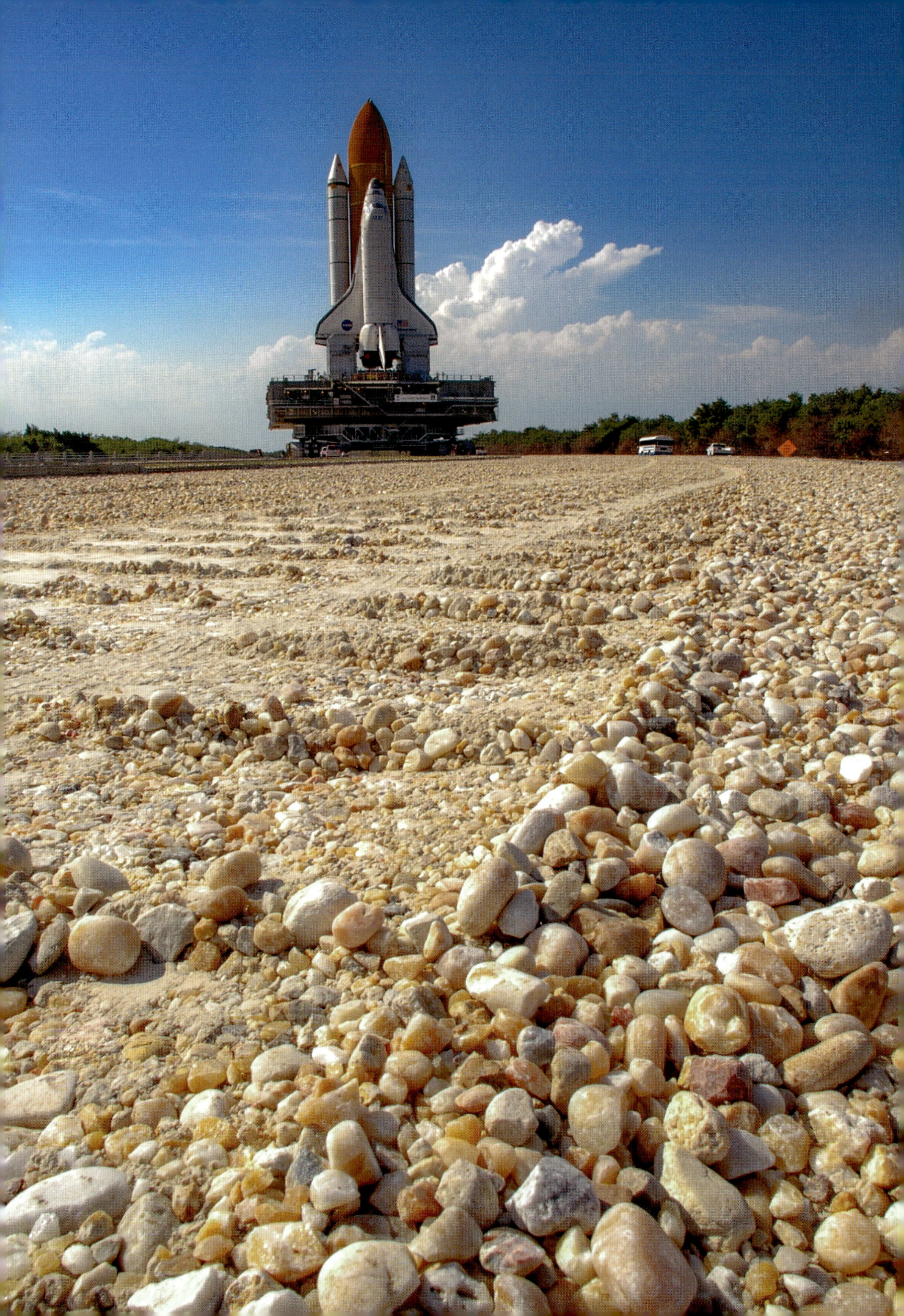

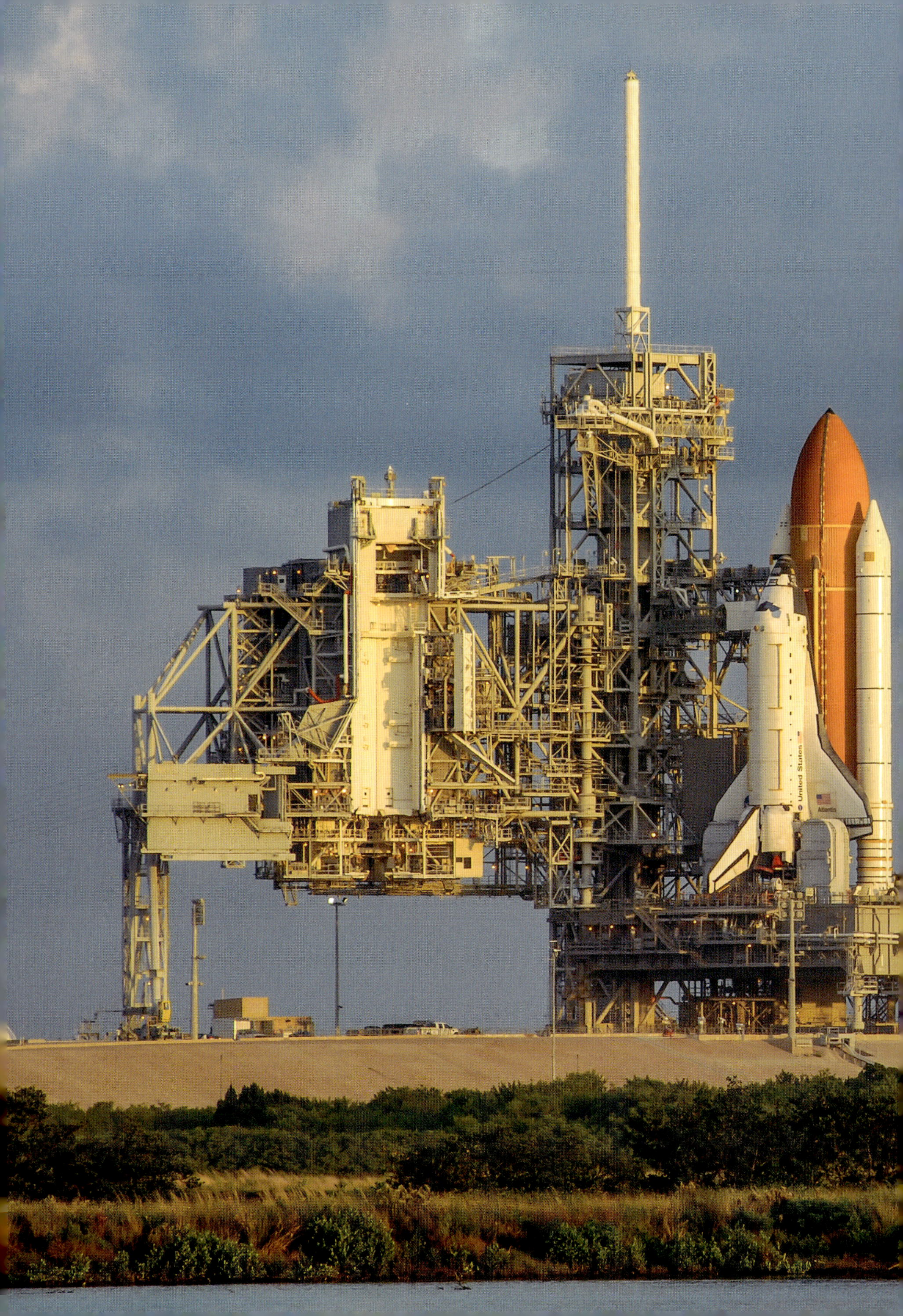
United States

A Rare Occasion

On rare occasion, two shuttles could be on the pad at the same time. On even rarer occasion, only four times in its 30-year history, both shuttles would be visible with the launch tower rolled back. In September 2018, with gorgeous sunrise light, it happened for the third time. Here, Atlantis sits on pad 39A (foreground), readying for the final mission to service the Hubble space telescope, while Endeavour, on pad 39B, will act as a rescue vehicle in case of an emergency. This plan was enacted after the 2003 Columbia accident for this, the only subsequent mission not to go to the space station.

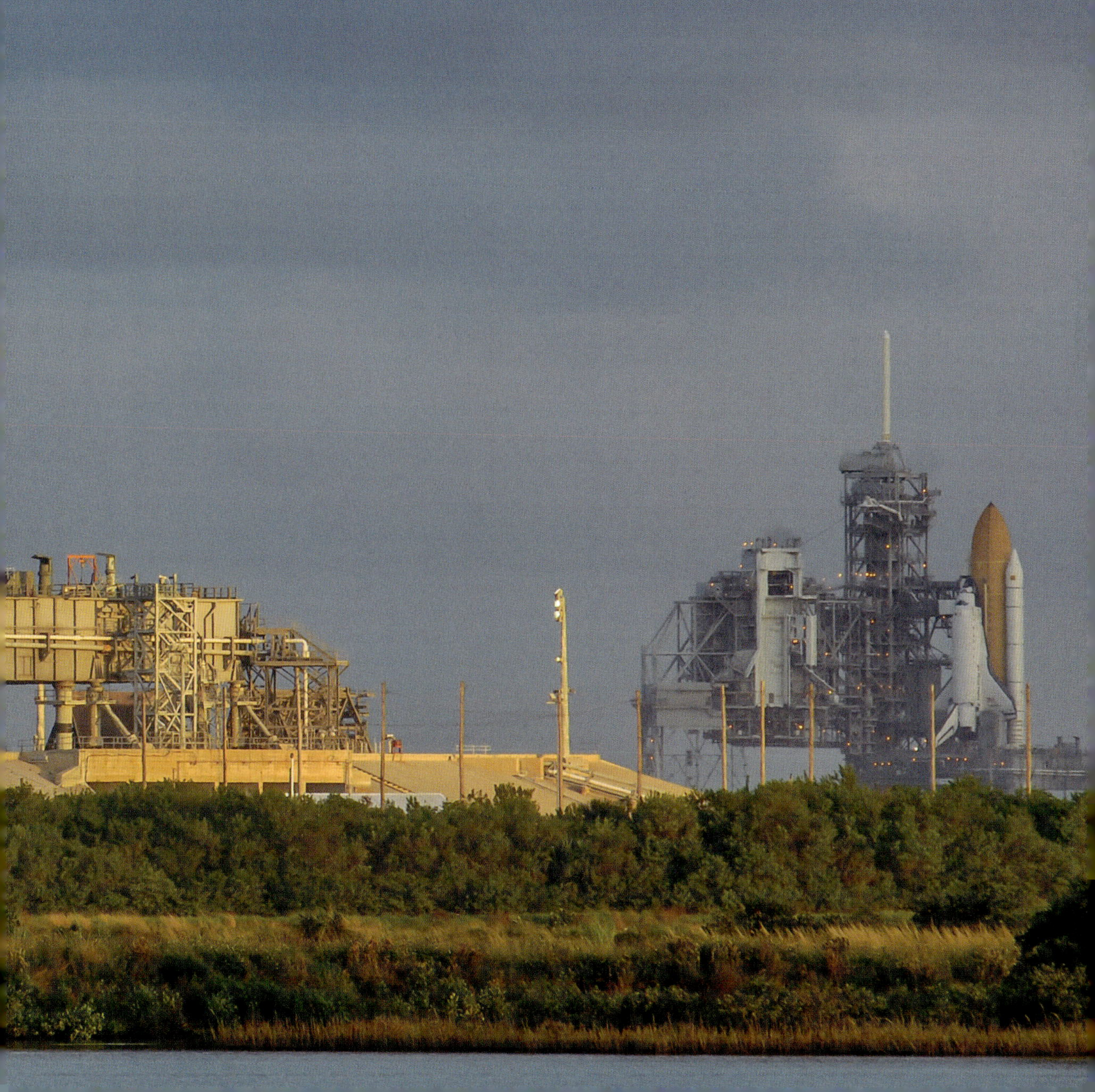

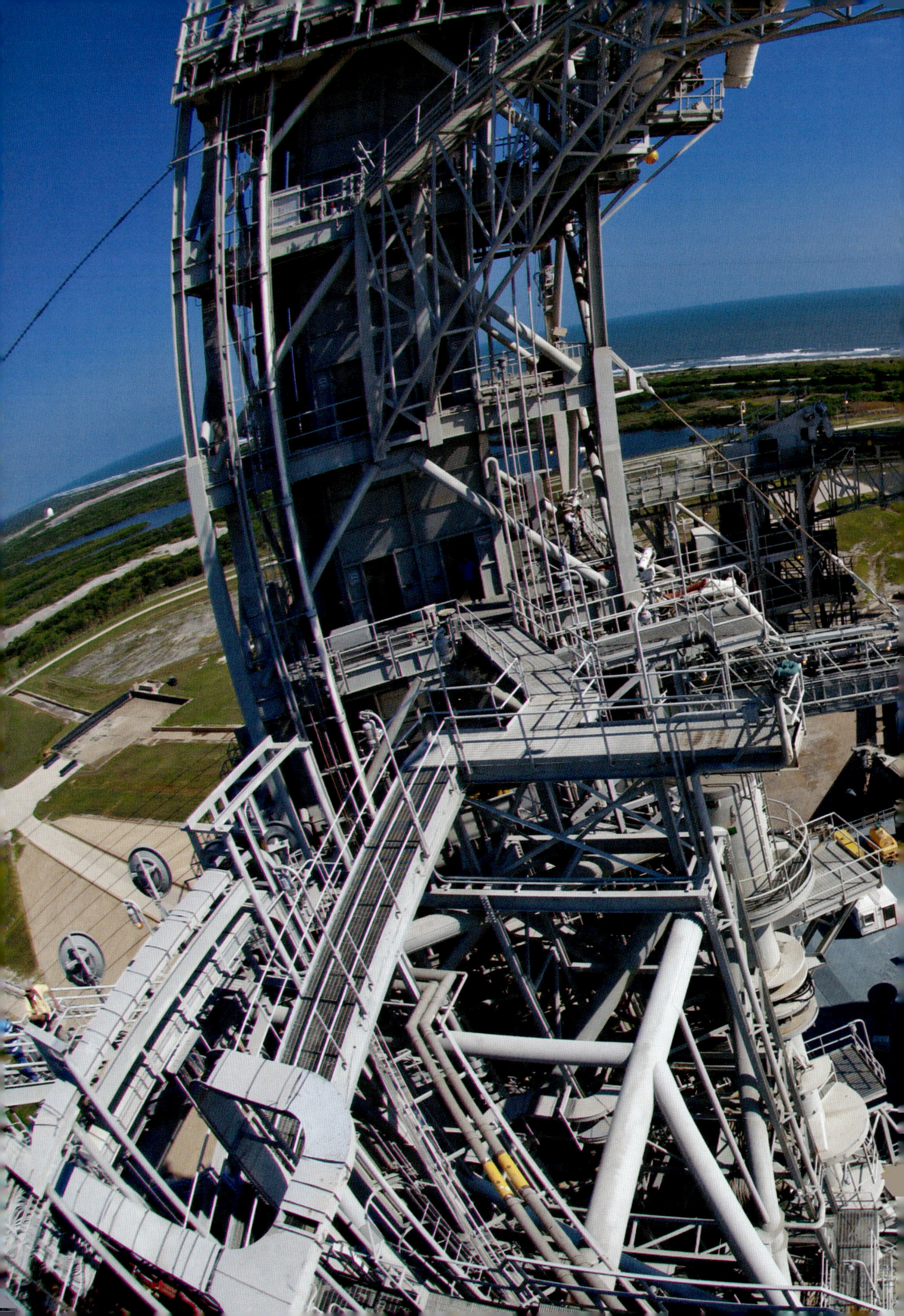

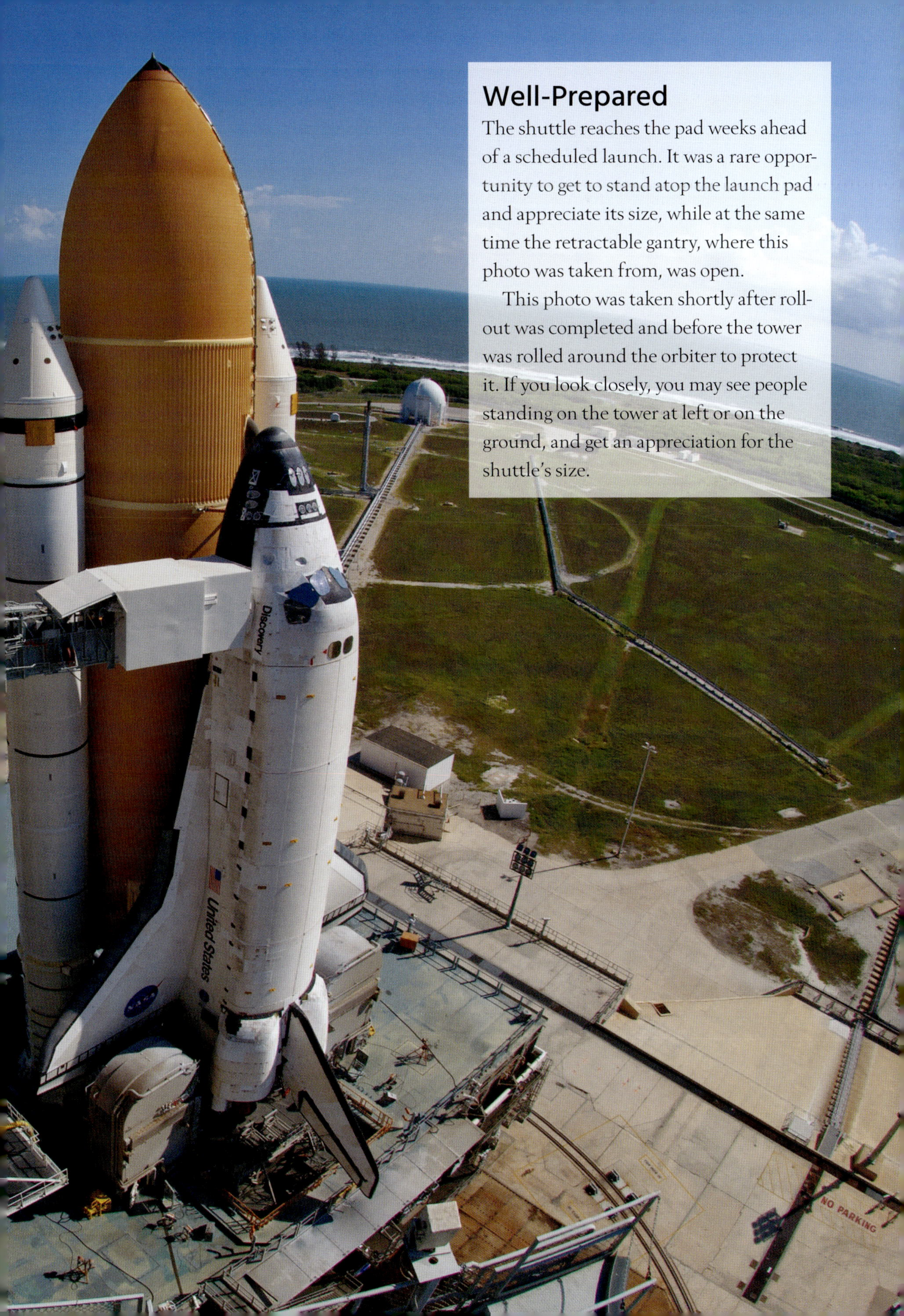

Well-Prepared

The shuttle reaches the pad weeks ahead of a scheduled launch. It was a rare opportunity to get to stand atop the launch pad and appreciate its size, while at the same time the retractable gantry, where this photo was taken from, was open.

This photo was taken shortly after roll-out was completed and before the tower was rolled around the orbiter to protect it. If you look closely, you may see people standing on the tower at left or on the ground, and get an appreciation for the shuttle's size.

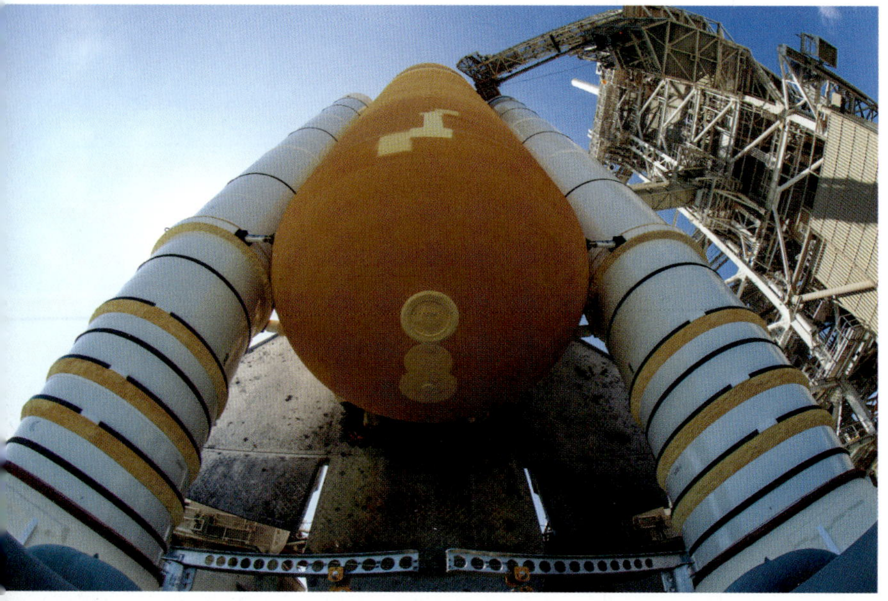

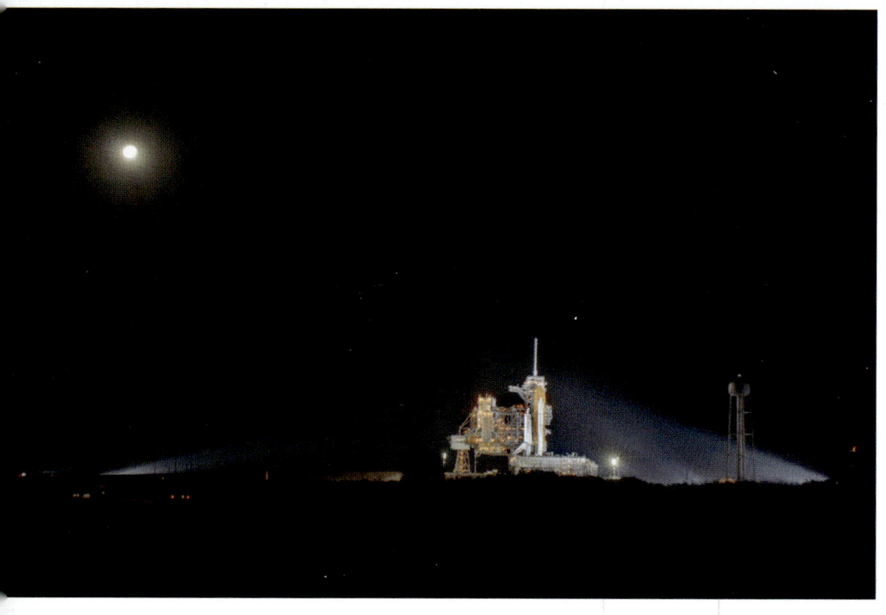

Mobile Launch Platform

(top) Standing under the shuttle stack on the MLP, a fish-eye lens is needed to take in the entire scene above.

Under the Light of the Moon

(bottom) The full moon sets behind the xenon-beam-lit shuttle Discovery the morning of the launch of STS-119.

At Sunset

(following page, top) Watching the sun set behind the shuttle the night before launch was common for launches that took place at midday or in the afternoon. As darkness set in, technicians would turn on the extremely bright xenon spotlights that were set up around the pad, normally placed at a distance of about 1,500 feet from the shuttle.

Reflections

(following page, bottom) The night before the last space shuttle launch in 2011, heavy rains made for some nice reflections around the pad.

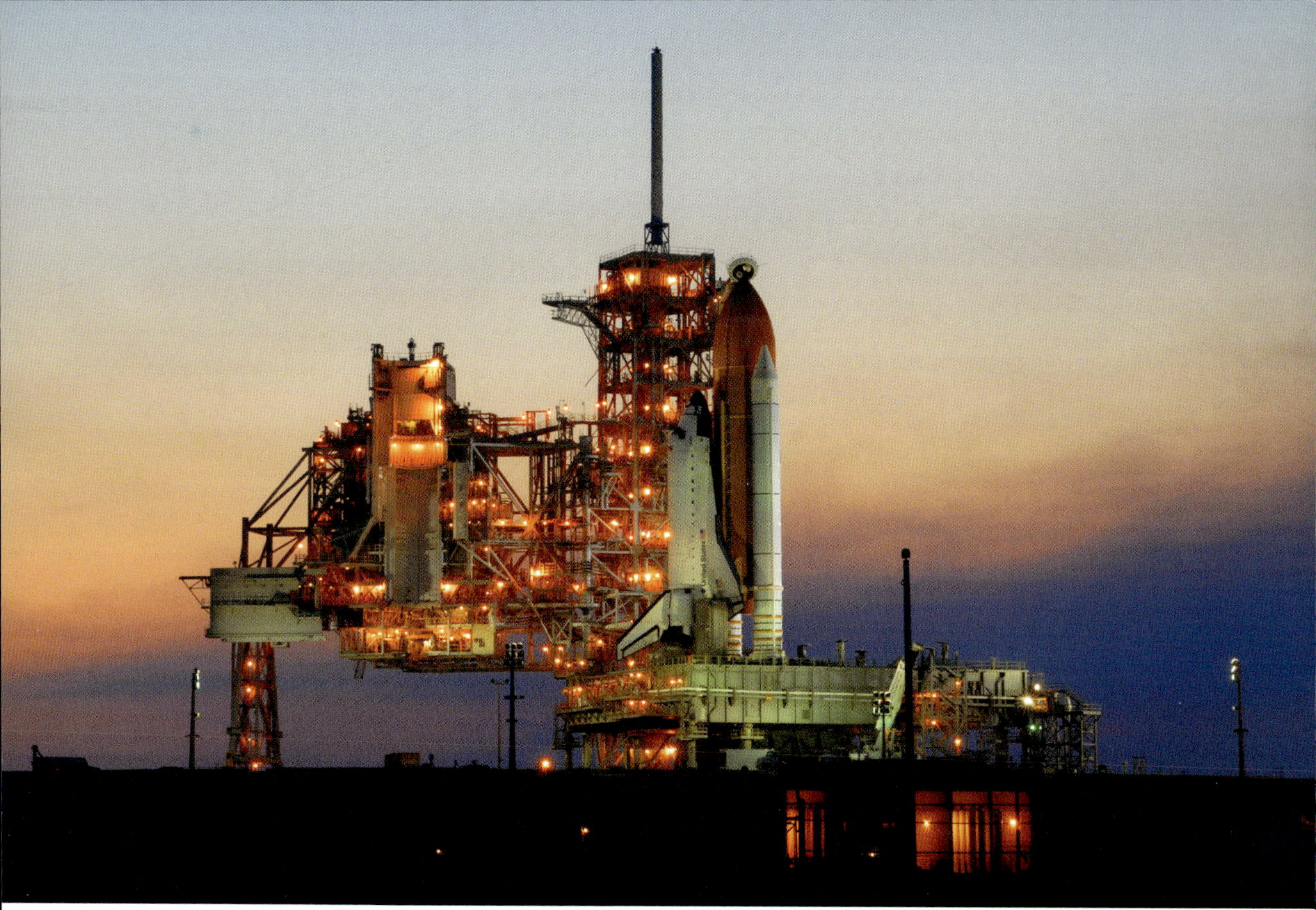

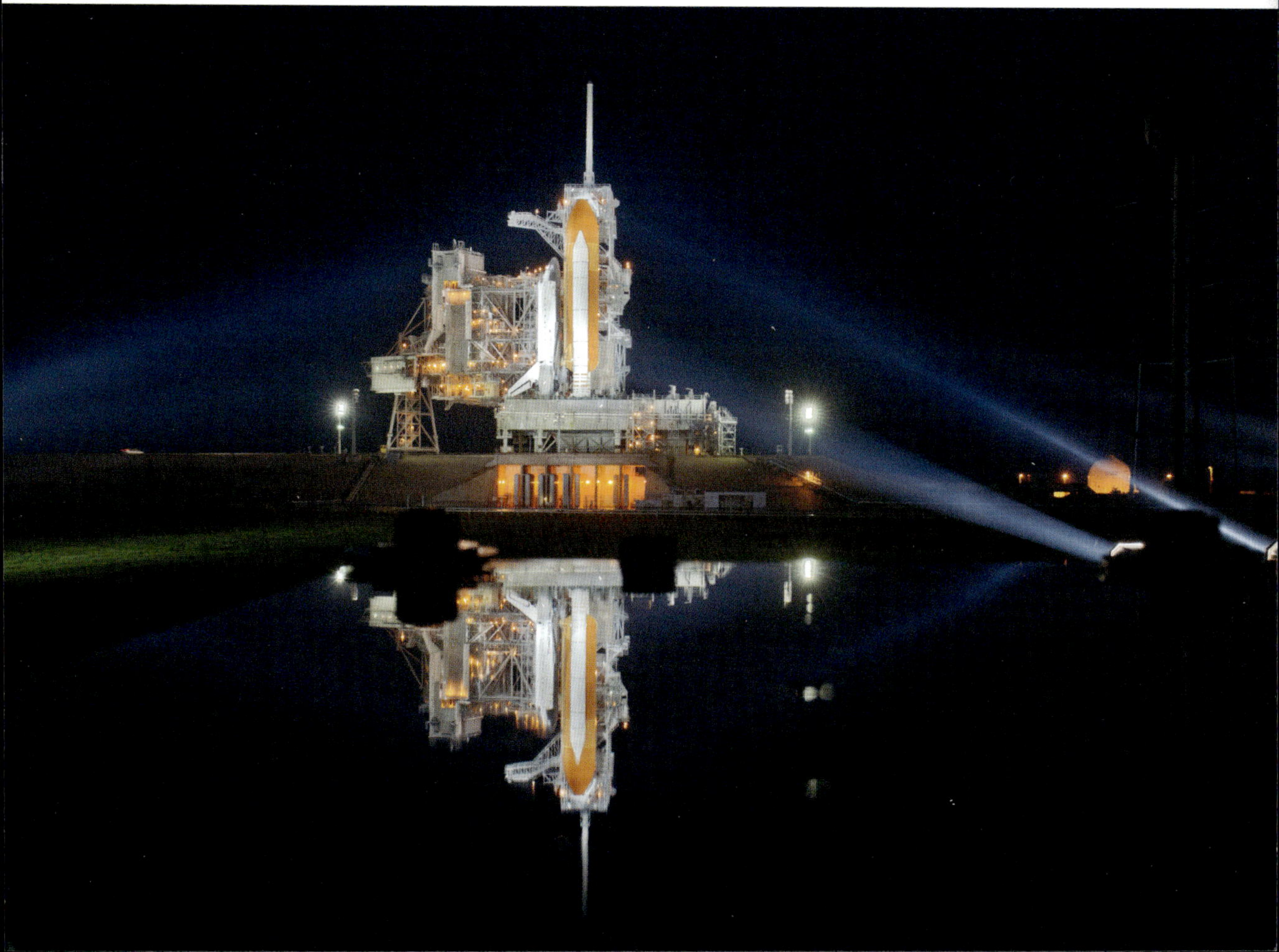

Personal Favorite

This is one of my favorite shuttle photos. It was the evening before the last space shuttle launch in 2011, STS-135. It was raining most of the day, and when they began to turn on the powerful xenon spotlights at sunset, the beams became very visible due to the moisture in the air. Scattered across the lawn in the foreground are many of the remotely placed cameras that will record the shuttle launch the next day.

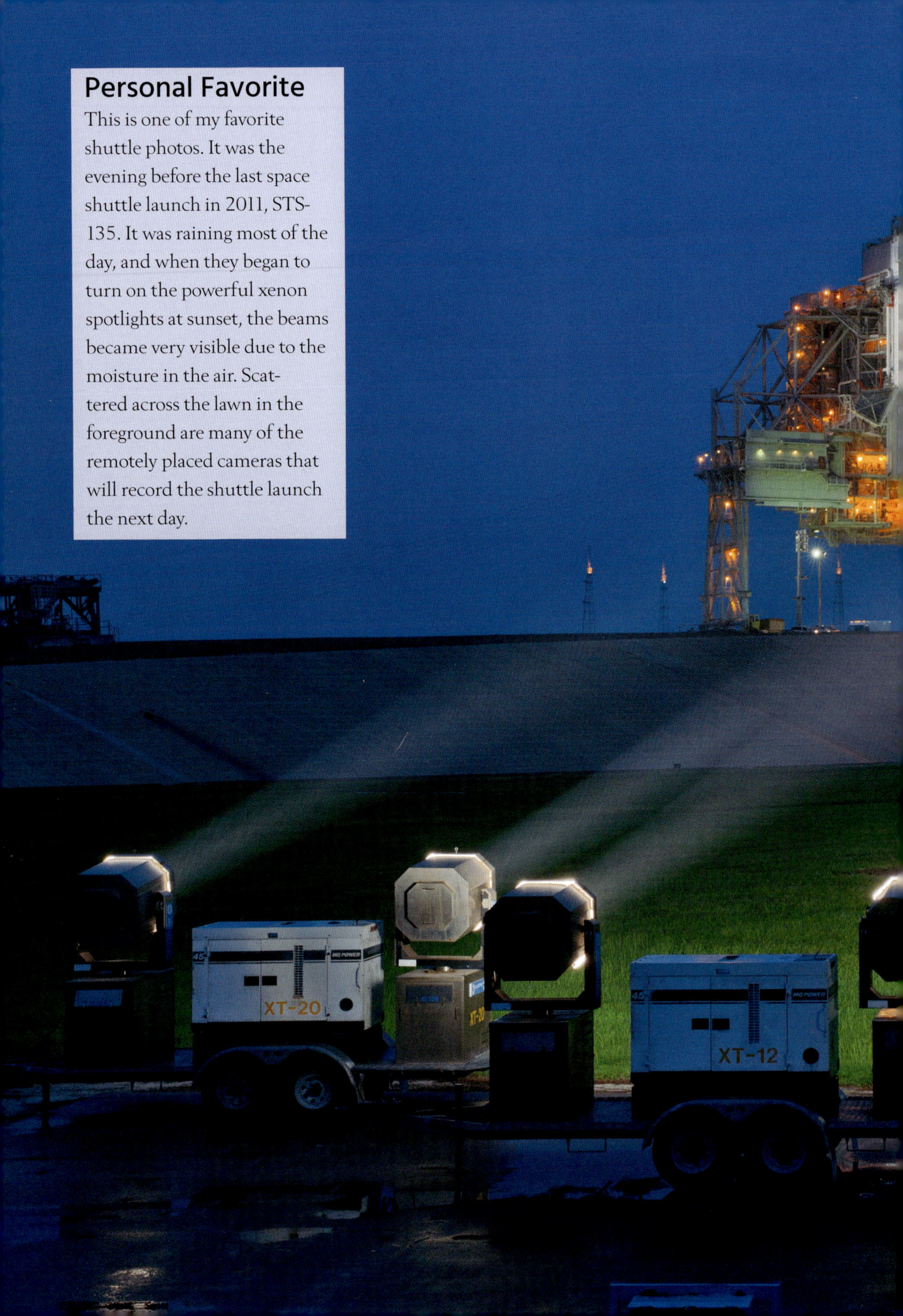

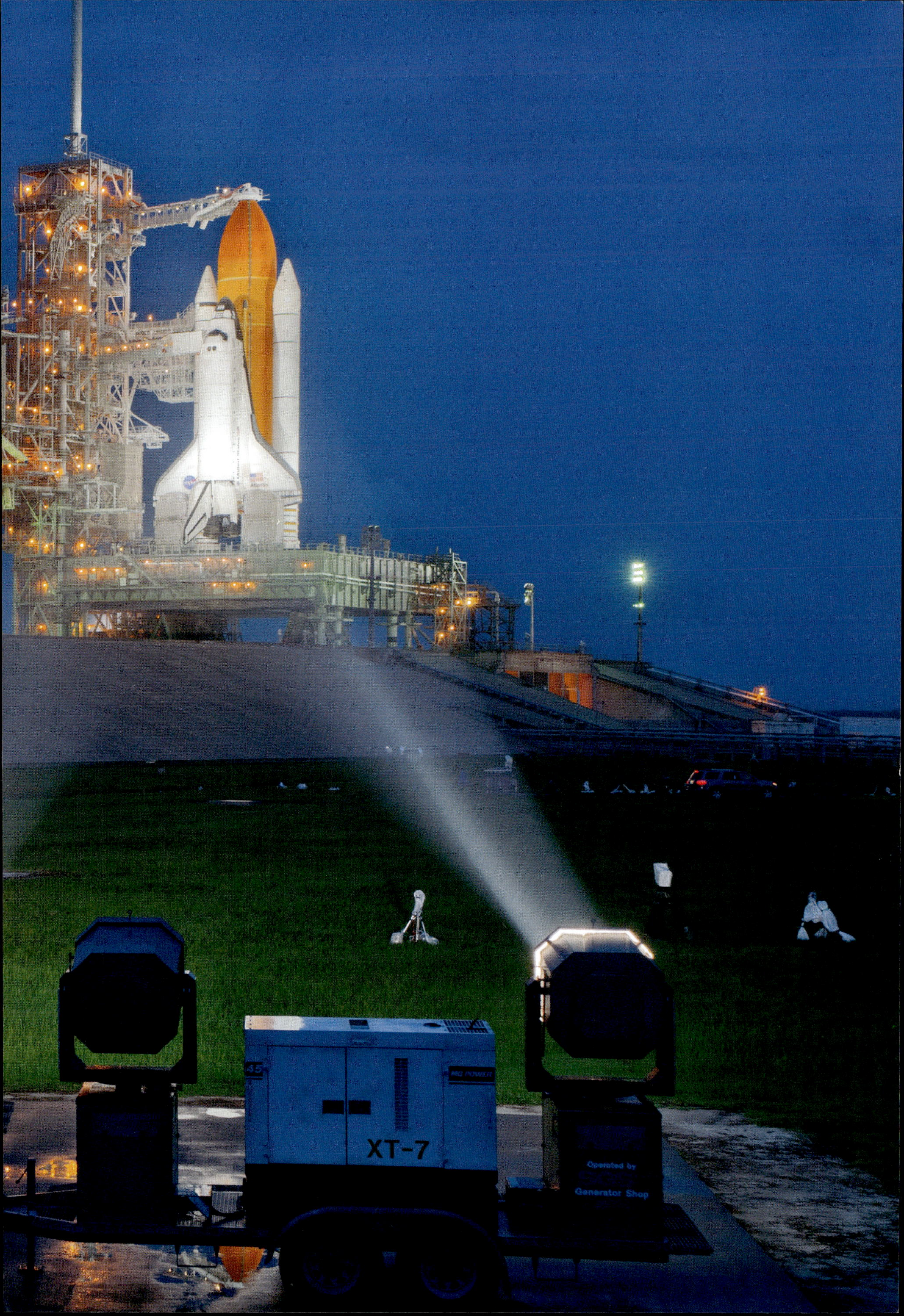
XT-7
Operated by
Generator Shop

Remote Cameras

The most challenging and unique part of photographing a shuttle or rocket launch is working with "remote" cameras (i.e., the launch pad cameras), which are set up ahead of time in locations that will not be accessible on launch day for safety reasons. Because no one is allowed closer than a couple of miles or more to a launch (it varies for each launch site and rocket), photos taken on or near the launch pad must be triggered on their own. That means you only get one shot to get it right: from the point the camera is left alone, which could be several hours to a full day ahead of time, you have to have the right settings in place, the camera focused and locked so that it will not move, and have nothing trigger it other than the launch in most cases. There is no do-over, and you have to be able to "see" the photo before it's taken; sometimes, there isn't even a rocket in sight to frame at the time.

A day or two before a launch, a suite of camera gear is loaded into the car or truck and taken out to the pad to be set up. For me, a typical launch setup could be 10 to 15 cameras, though I have used as many as 35 for a launch to date.

One by one, each camera is placed in a desired spot; the whole setup can easily take the better part of a day, or even be done over two days for some launches, depending on the number of cameras used and how long it takes to place each one. Camera locations can be a mile or more apart from each other at times.

Tripods are secured to the ground when possible, using things like tent stakes, or even sandbags. Other times, specialty clamps can be used in place of tripods, such as on elevated towers where tripods would not be secured well

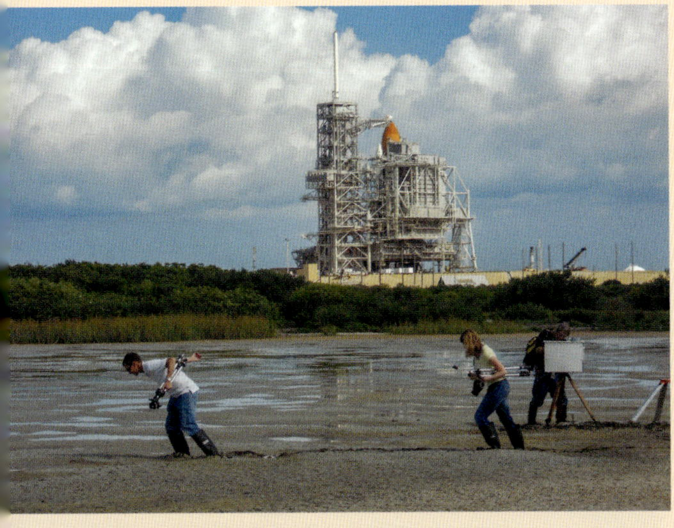

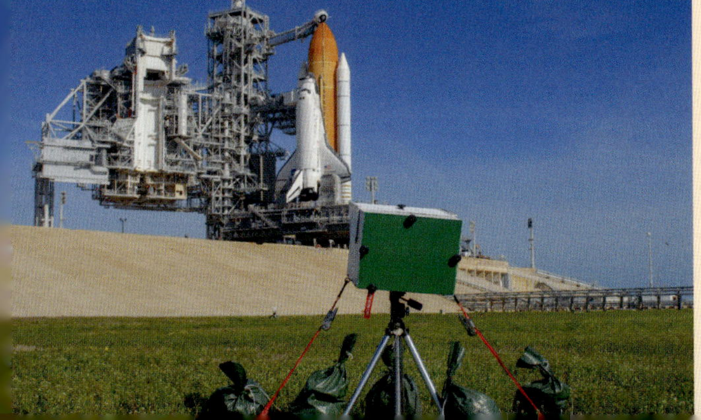

Setup

(top photo courtesy of Craig Durling) Photographers sometimes trudge through the swamps, frequently through quicksand-like mud, to set up unique shots ahead of a shuttle launch.

Weather-Proof Housings

(bottom) For their protection, cameras are frequently enclosed in weatherproof housings such as this one, or at the very least covered and secured with plastic bags with only the lens exposed to protect them from the rain. Here, due to its close proximity to the shuttle, the camera is also sandbagged to lower the risk of any potential flying debris as a result of the powerful blast.

enough. Safety tethers made of steel normally are also incorporated to prevent cameras from falling or being blown away by the launch. Safety is a top priority in the space program, and any camera gear used must be secure.

Just as importantly, all gear must be protected from the elements: rain is often a factor in Florida, but so are things like salt air, and the exhaust from the rocket launch itself, which can be instantly corrosive in some cases.

The cameras must have a way to be triggered on their own. Most of the time, sound is used as the primary trigger: when the rocket engines ignite, the camera fires. (Sound levels from the launch in the immediate vicinity can reach levels that would make a person permanently deaf in an instant.) Timers are used in some cases when sound may not reach soon enough, or the camera is too far to trigger in time. Other photographers have been known to use vibration or light from the intense flames to set cameras off.

Launches can do a number on camera gear, especially when close. The incredible sound, heat, vibrations, and launch exhaust can wreak havoc if the equipment is not protected well enough (and even when it is). During a launch, it is common for dirt, sand, and pebbles, and other debris to be sent flying at high speed away from the area around the rocket, and into the lenses or whatever is protecting them. Thin glass can be shattered.

Setups can be fun, but can also come with a tiny bit of danger. I have trudged through swamps with quicksand-like mud, been cut by sharp branches in the mangroves, and alligators and snakes are common in what is, in fact, a national wildlife refuge. Perhaps the worst things, however, are the humidity and mosquitoes.

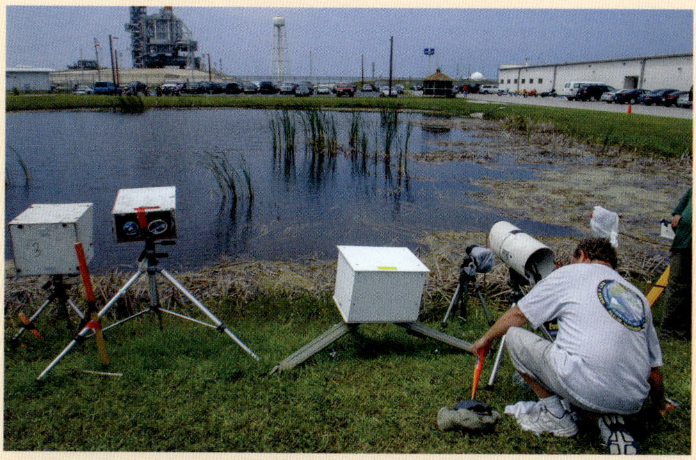

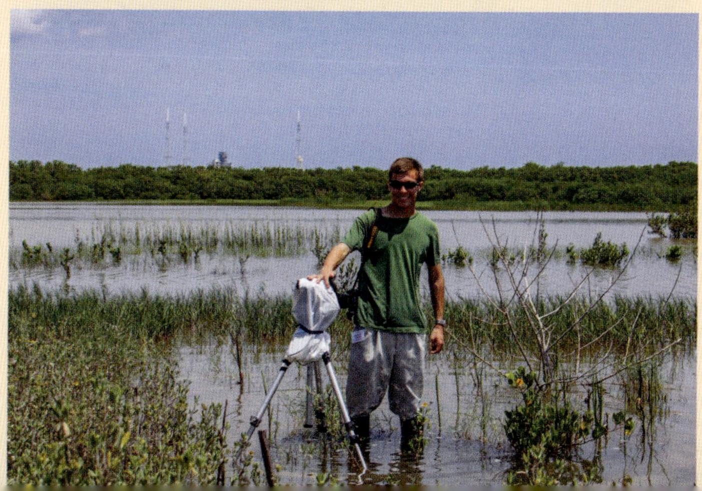

Final Check

(top) Photographer William G. Hartenstein checks a remote camera at a popular media spot ahead of a shuttle launch in 2005.

Yours Truly

(bottom photo courtesy of Wiliam G. Hartenstein) A photo of me in the swamp at an shuttle camera setup.

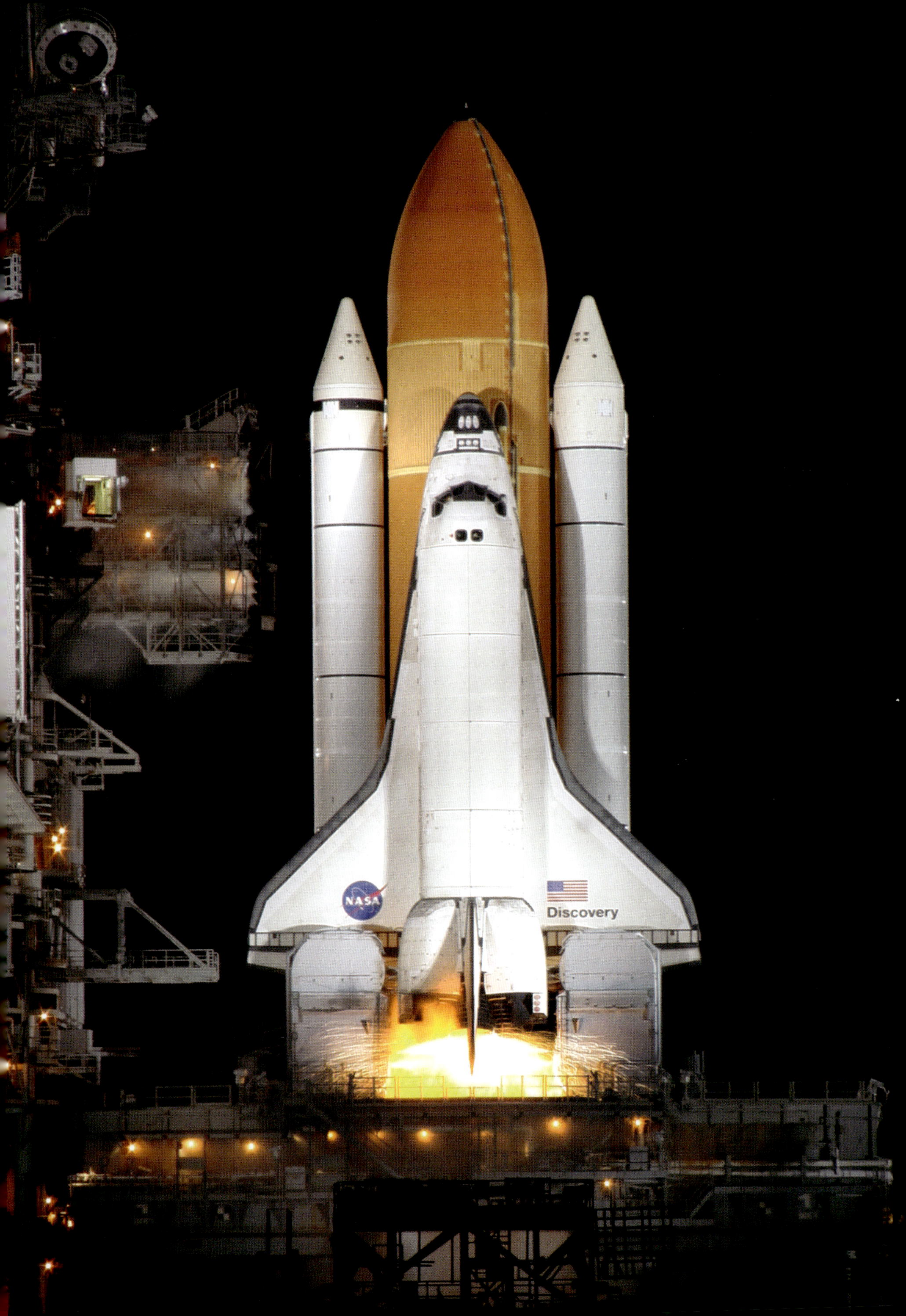
NASA
Discovery

LIFTOFF

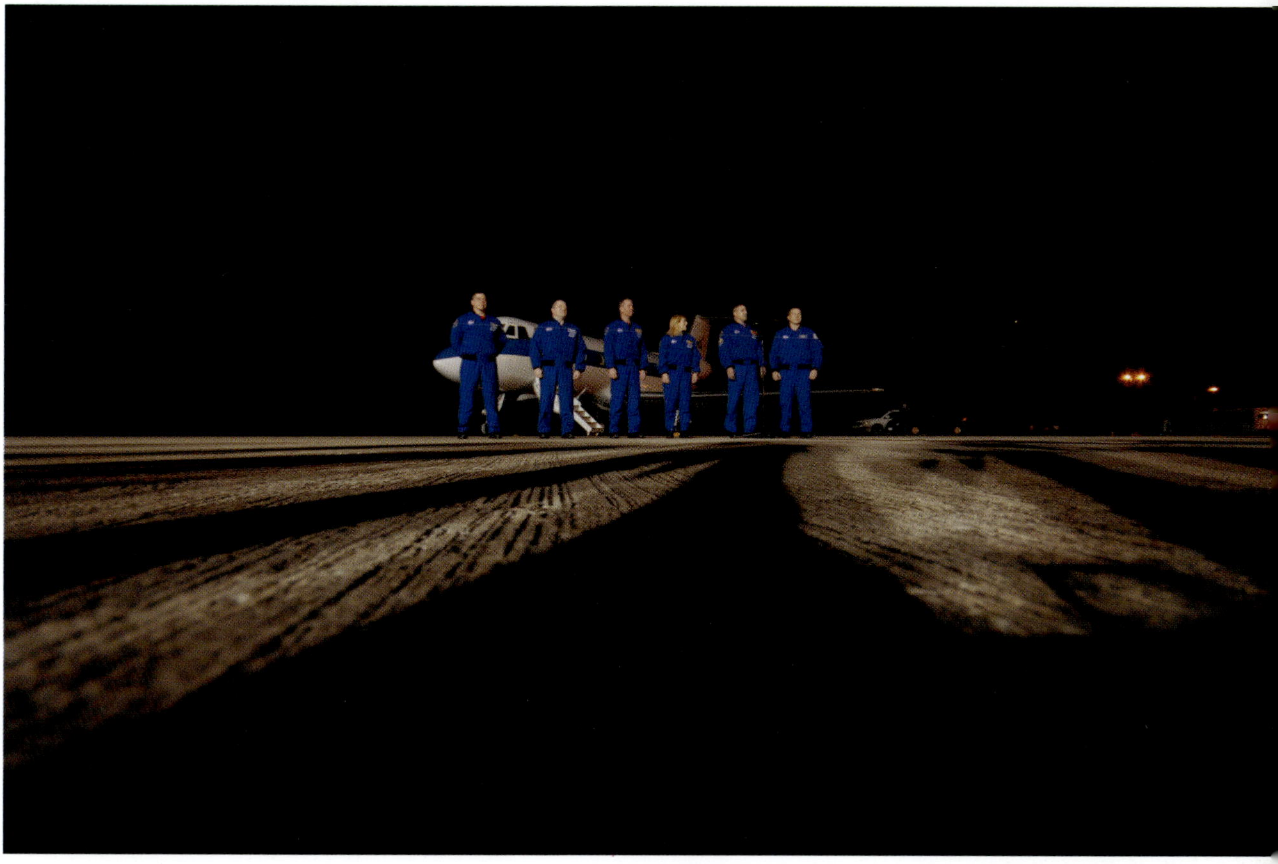

Planning

(previous page) Right before the engines ignite, a sparkler system was used to burn off any excess hydrogen gas lingering near the base of the shuttle to prevent an explosion. This photo was taken just a few seconds before main engine ignition of STS-116.

People

(above) I've never been quite as focused on capturing unusual shots of the astronauts or people, but for the nighttime arrival of the STS-130 shuttle crew a few days before their launch, I decided to try placing my camera on the concrete taxiway, while members of the media around me focused on their head shots and questions. Lights were set up behind the crowd.

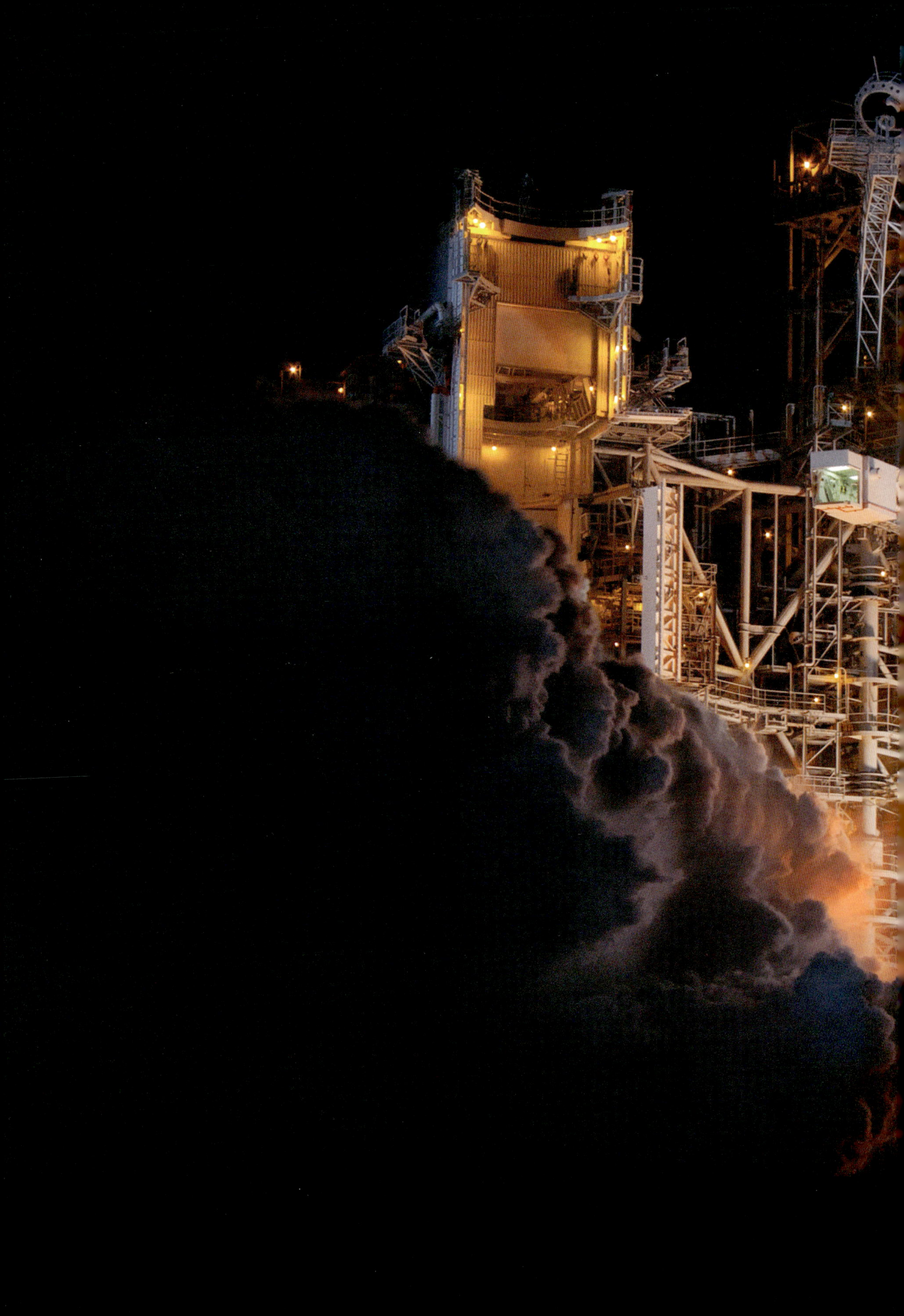

Engines Ignited

The shuttle was an incredible machine, and amazing to photograph. Once cameras were in place around the pad, finished the day before, the area was cleared, fuel was pumped in, the astronauts boarded, and off it went. In this unusual view, the three main engines on Discovery have just ignited but the shuttle is still a few seconds away from launch, which doesn't happen until the two giant, solid rocket boosters ignite.

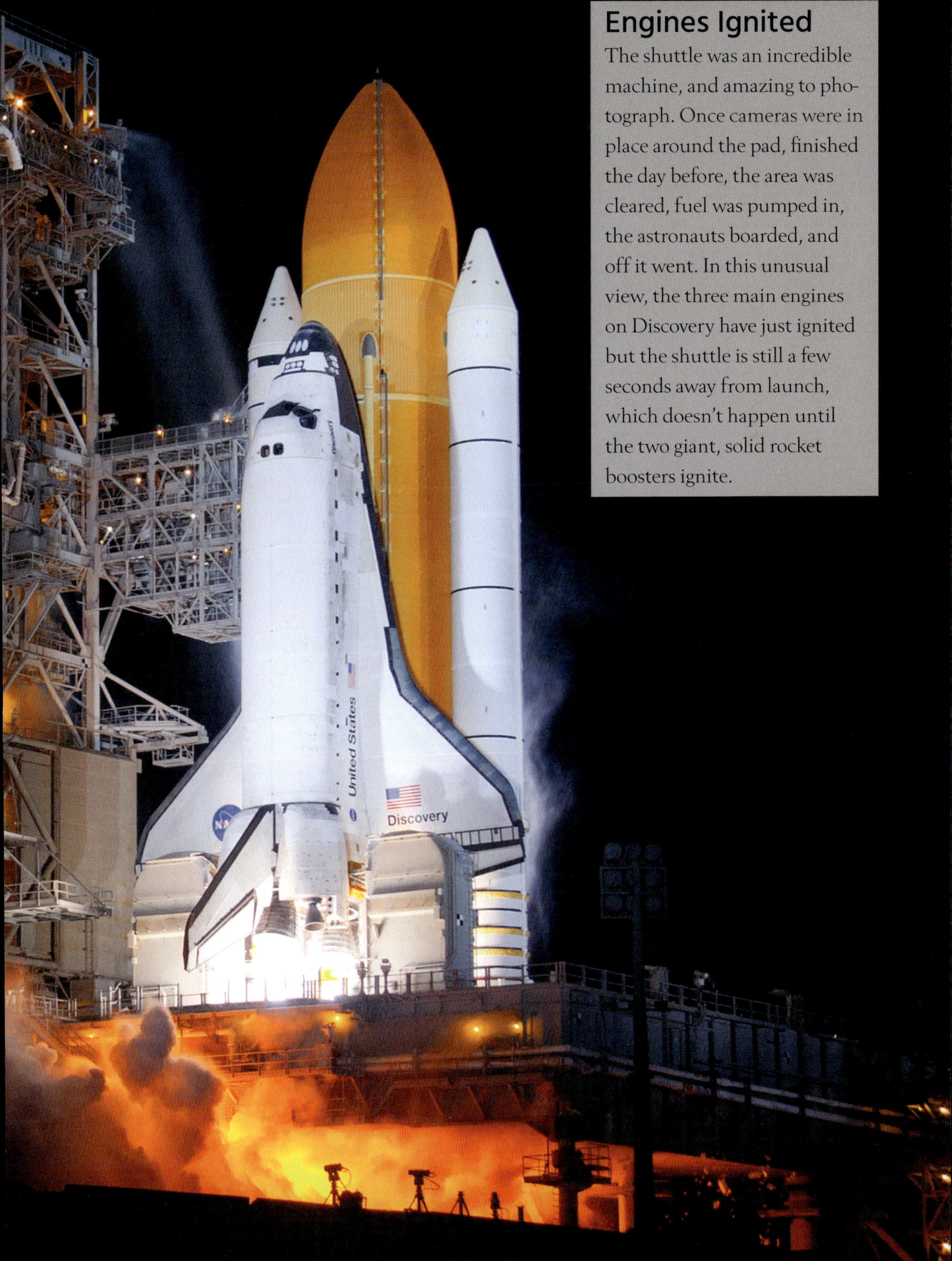

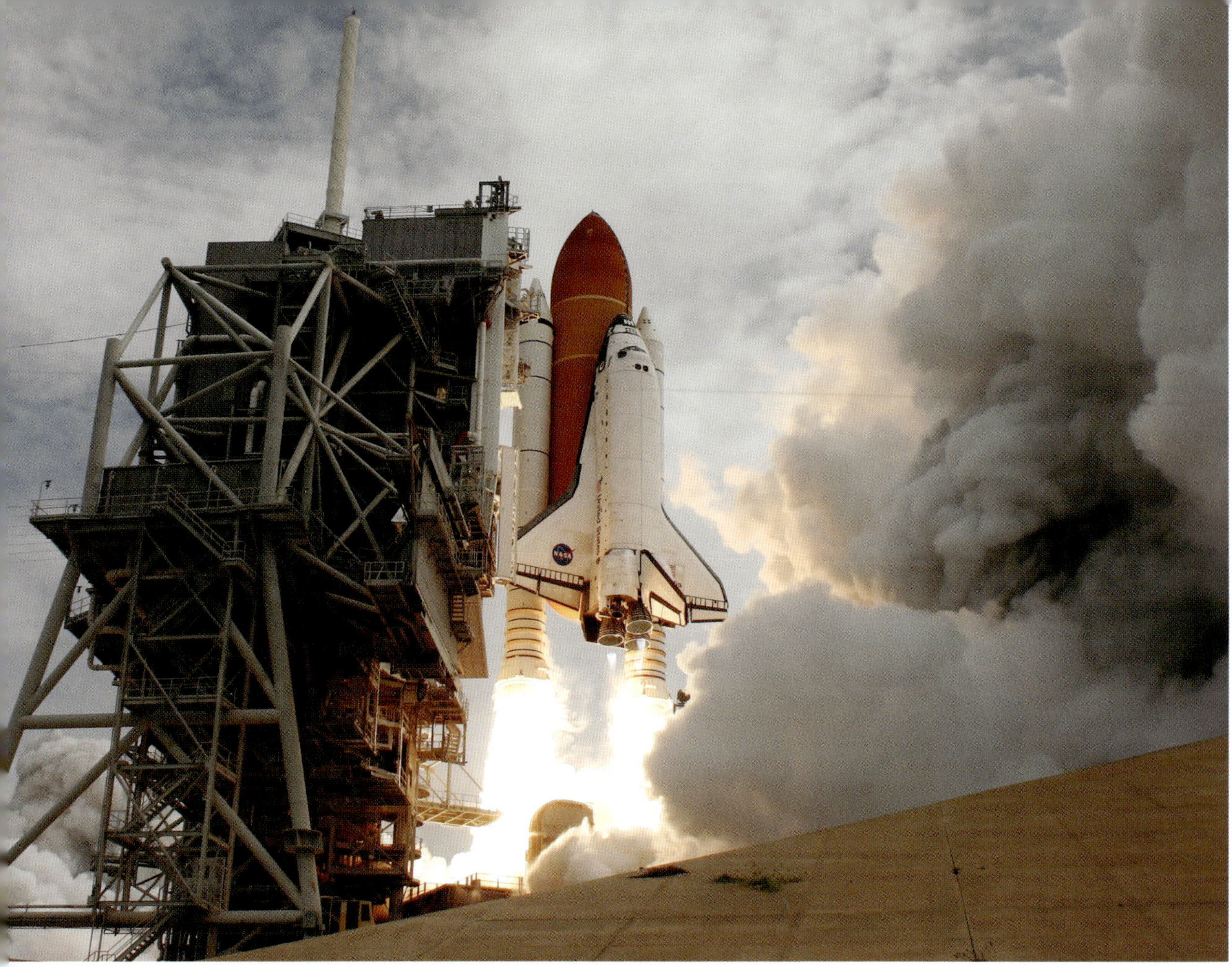

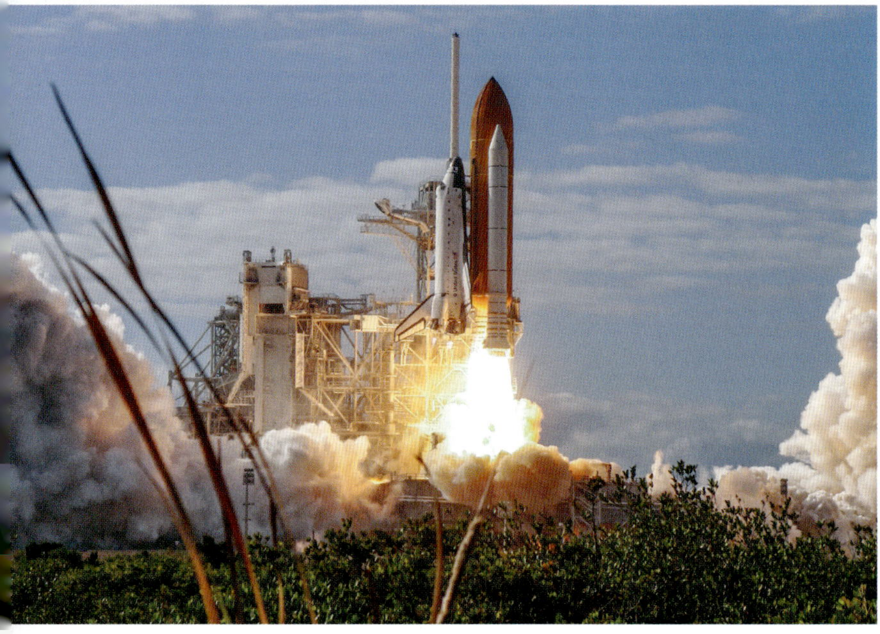

STS-135 Atlantis

(top) The twin solid rocket boosters ignite and propel the space shuttle off the pad, totaling, along with its three engines, nearly 7.8 million pounds of thrust. Here, Atlantis is launching on the final mission of the program, STS-135.

STS-129 Atlantis

(bottom and following page) Framing the launch with vegetation is a favorite compositional approach, and it's fairly easy to do in the wildlife refuge that surrounds the pad for miles.

Nature and Technology

Birds frequently wind up in the frame as well.

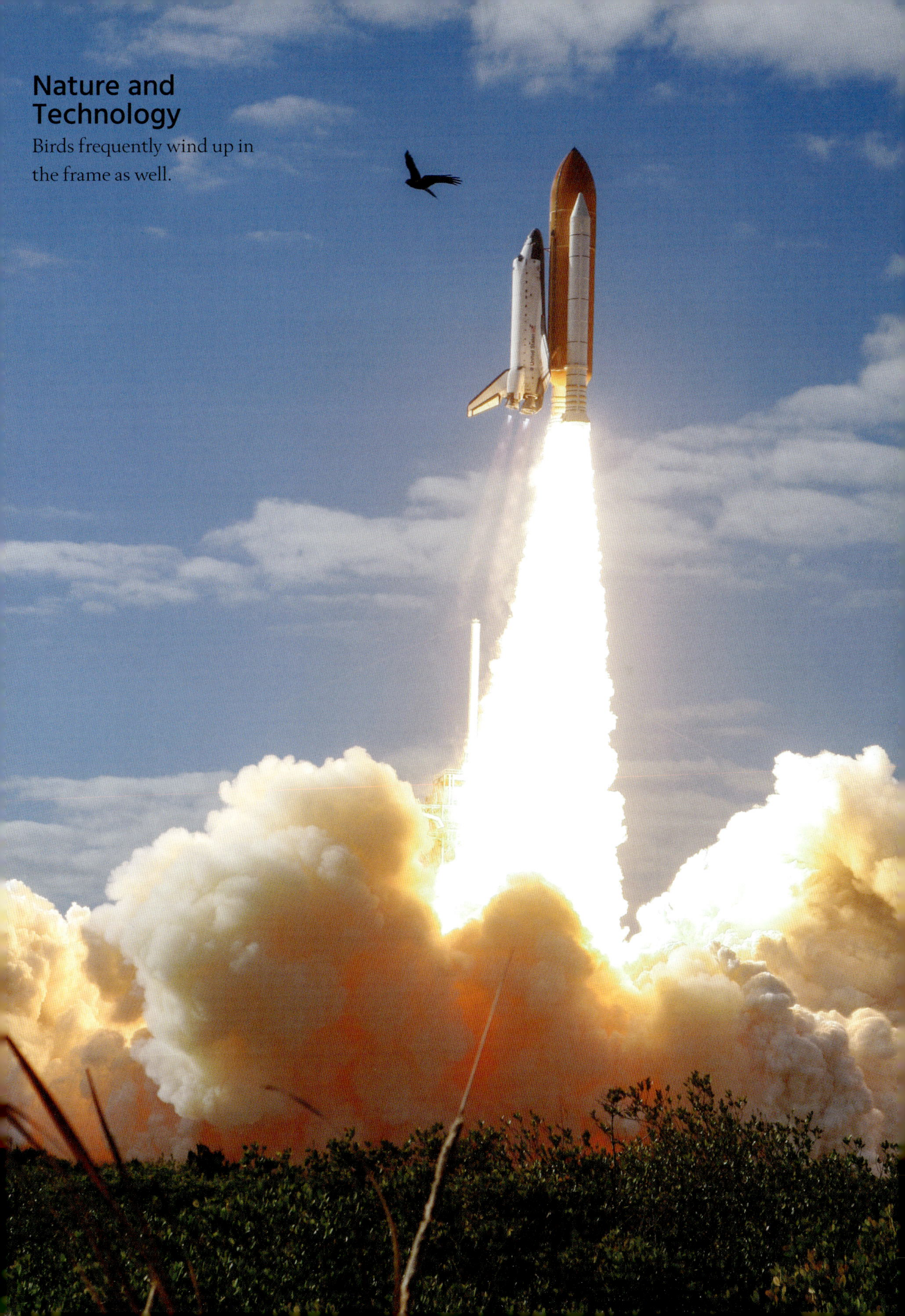

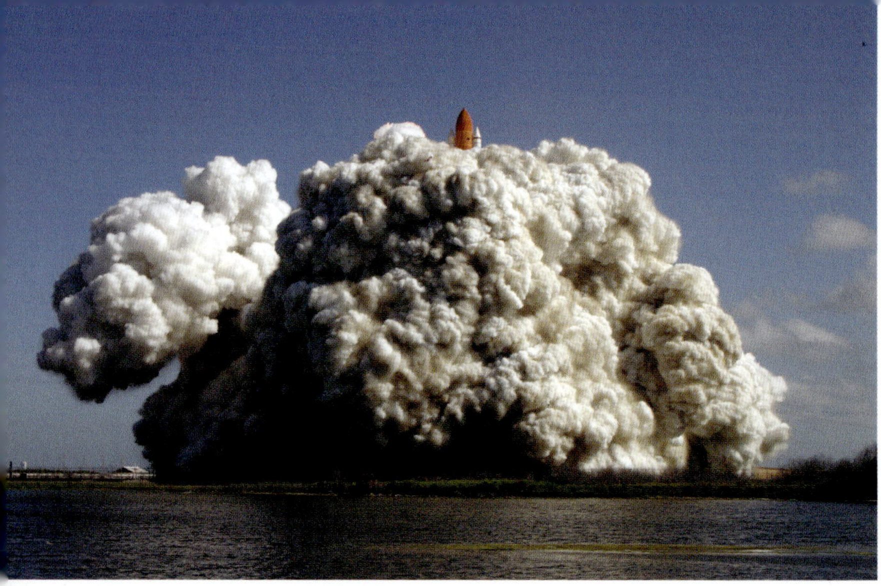

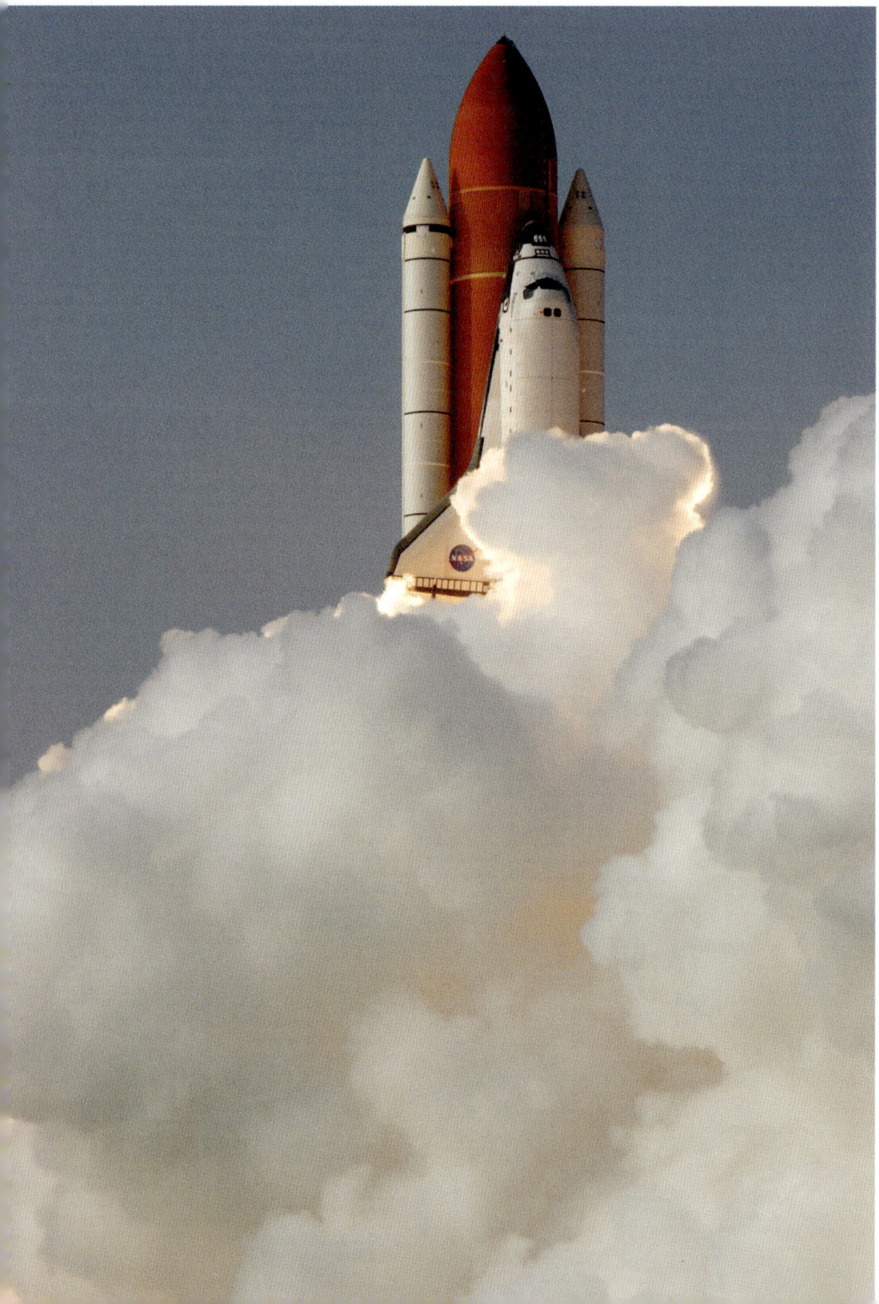

Smoke

(top and bottom) One great thing about photographing the space shuttle was the way the smoke behaved. The shuttle launch pad was designed to send it in opposite directions; when the three main engines lit, steam would be sent out to the south, building for six seconds before the solid boosters would ignite and have their smoke thrown to the north. This allowed for shots, though they could be risky, where the shuttle was only partially visible, its nose just sticking up as it rose, or streams of smoke and steam being blown by the wind left or right and partially blocking the view. (This was risky because you could miss the shot entirely on some launches.)

The first shot *(top)* was taken from a rarely used location due north of the launch pad's flame trench, STS-124. Discovery's orange fuel tank is just visible popping up from behind the solid fuel exhaust. In the second shot *(bottom)*, Endeavour is just visible, the NASA logo on its wing framed by exhaust, as it launches on STS-118 in 2007.

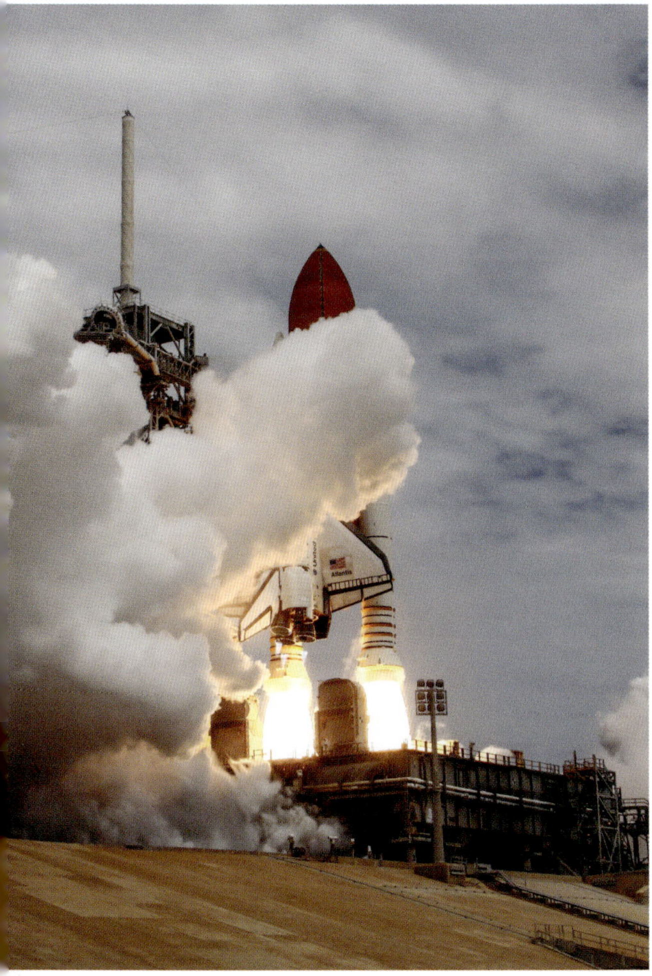

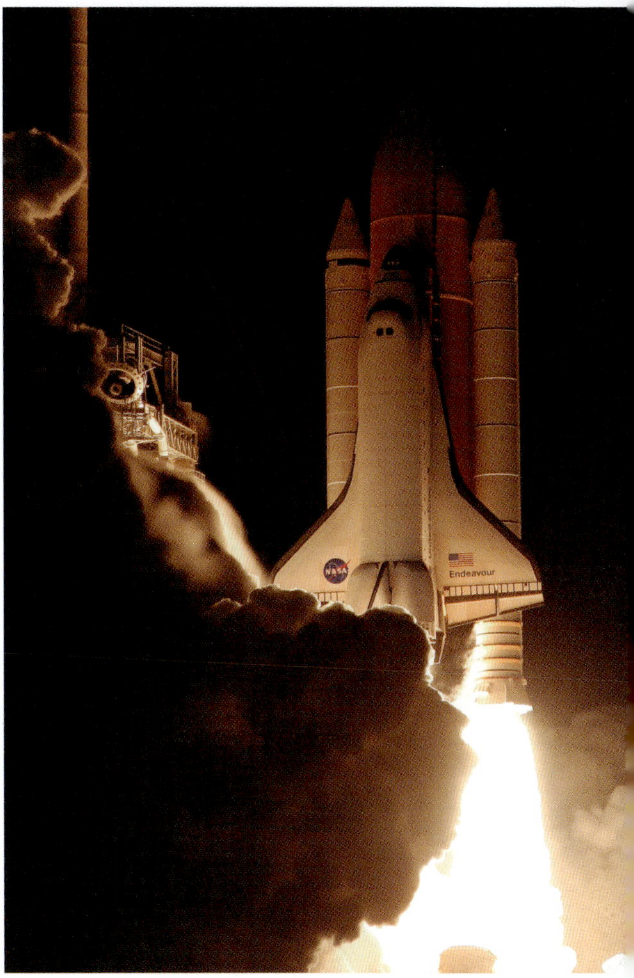

Unpredictability

(left) You could call this a perfect example of the unpredictable. An offshoot of steam blotted out Atlantis as it began its final journey into space in 2011 (STS-135). It's almost as if nature were going out of its way to ruin a shot! The result is amusing in ways, and unique.

An offshoot of steam has blotted out Atlantis as it begins its final journey into space . . .

Increased Drama

(right) At night, the stream and smoke can be even more dramatic, as the light from the engines casts a golden halo around the edges. Here, Endeavour rises on STS-123.

Endeavour's Final Mission

(following page spread) The shuttle Endeavour lifts off on its final mission, STS-134. Cloudy conditions made for dramatic light on this day in May, 2011.

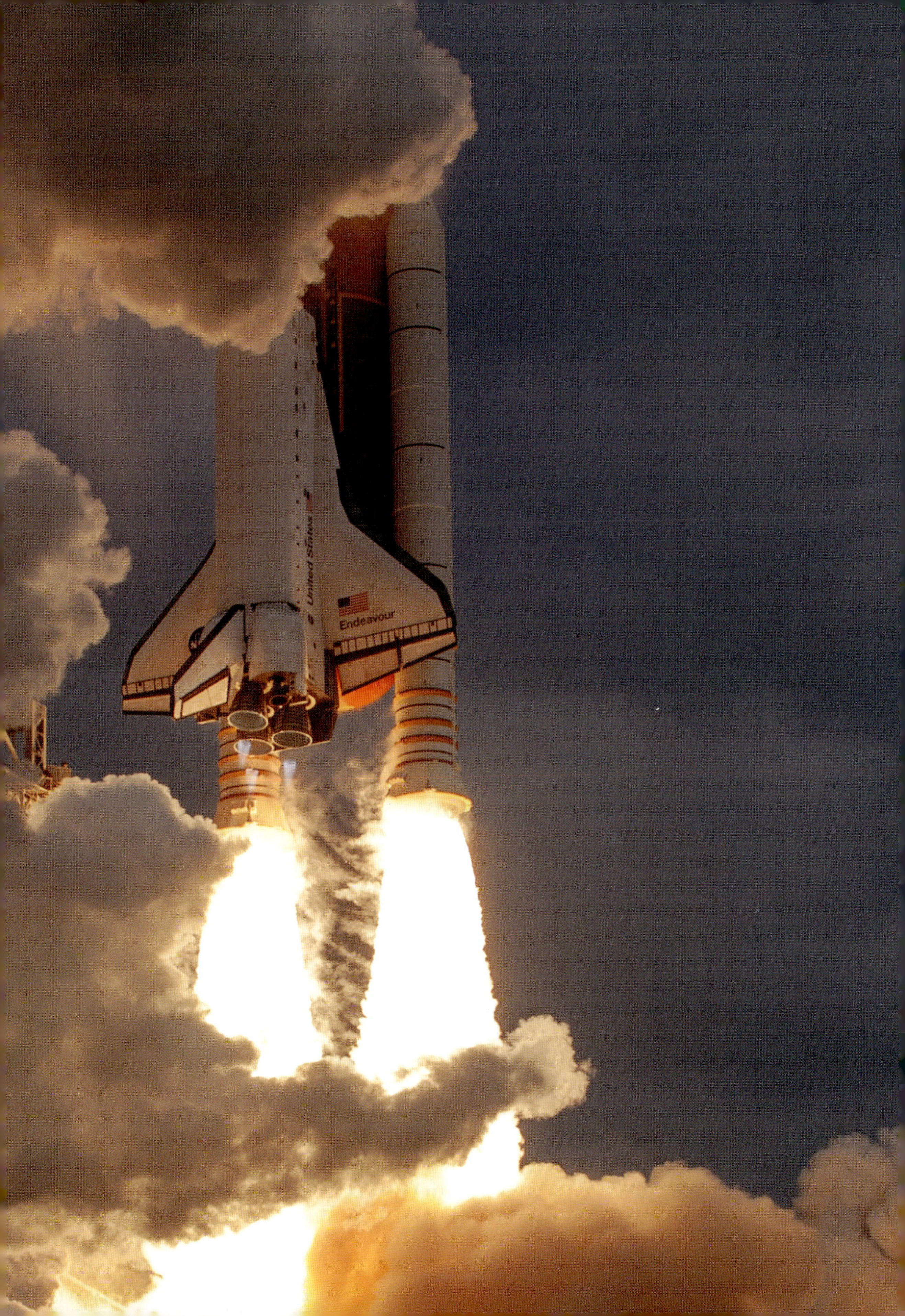
United States
Endeavour

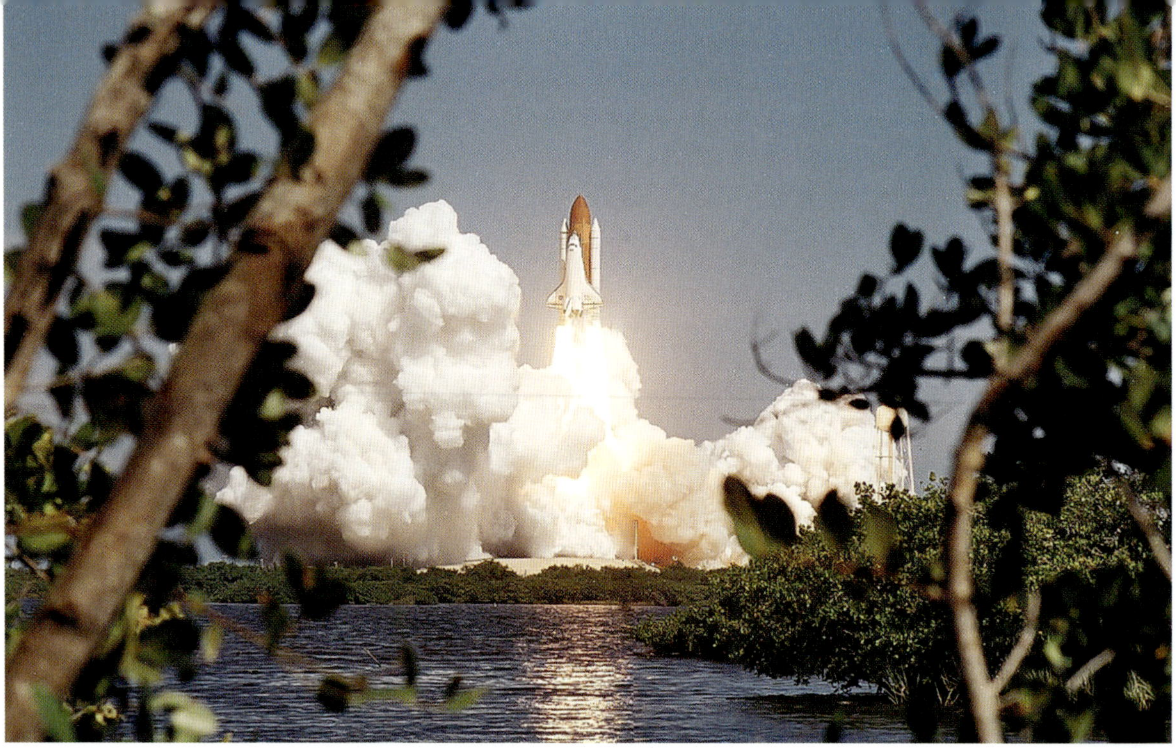

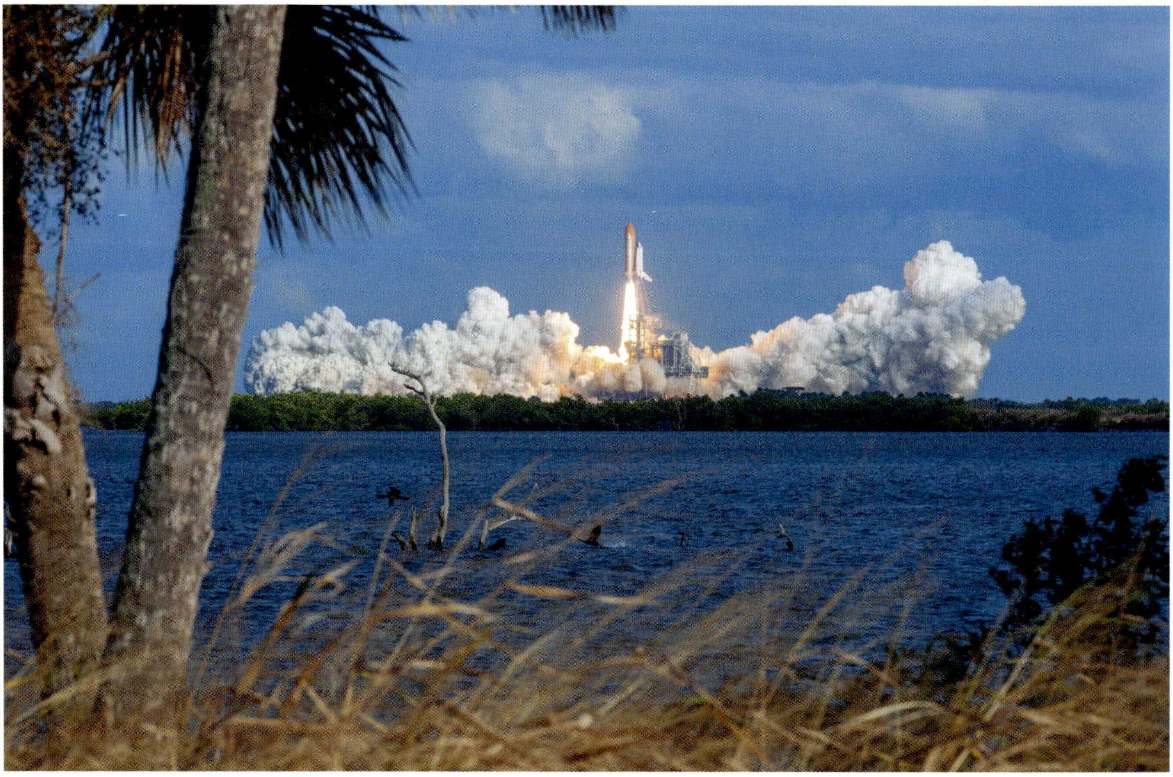

Framing

The surrounding wildlife refuge provides a wealth of opportunities to frame launches with trees, mangroves, logs, palm trees, and dead branches, and to get the occasional bird in the shot. In the top image, Atlantis launches on STS-115; in the bottom shot, on STS-129.

From the Lagoon

(top) Atlantis begins its trip on STS-115, as seen from this camera set in one of the lagoons around pad 39B.

(bottom) Discovery clears the launch tower on STS-120.

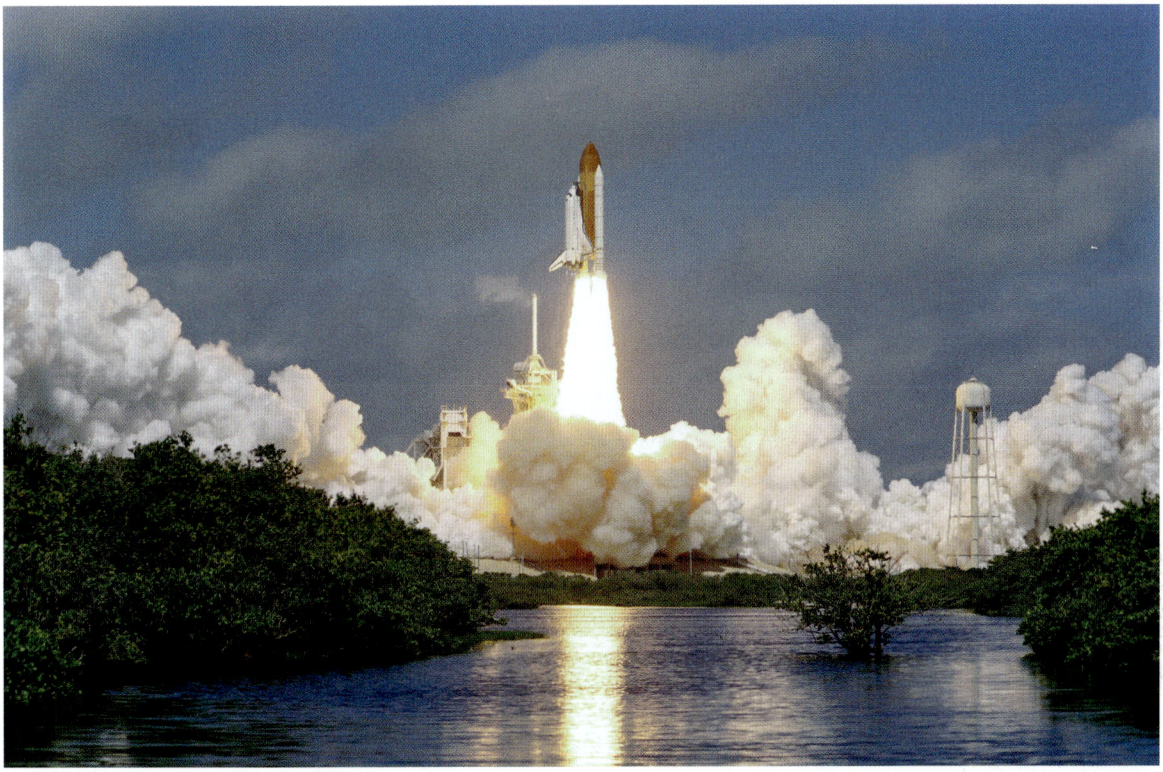

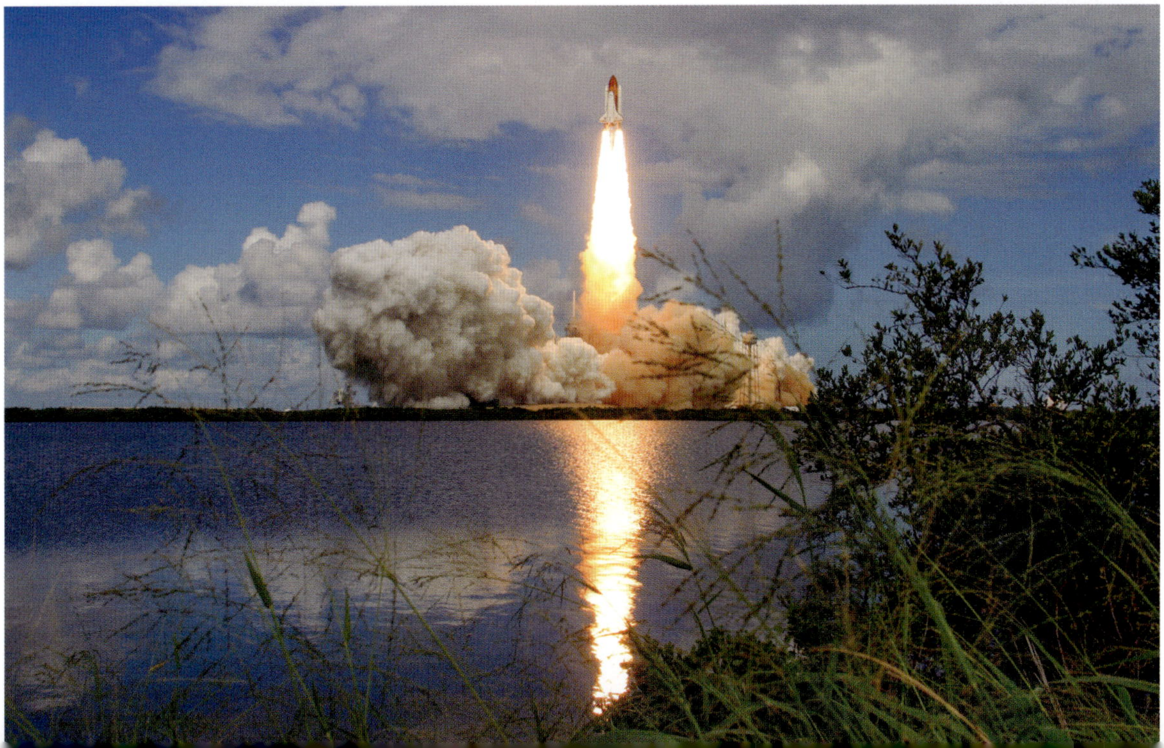

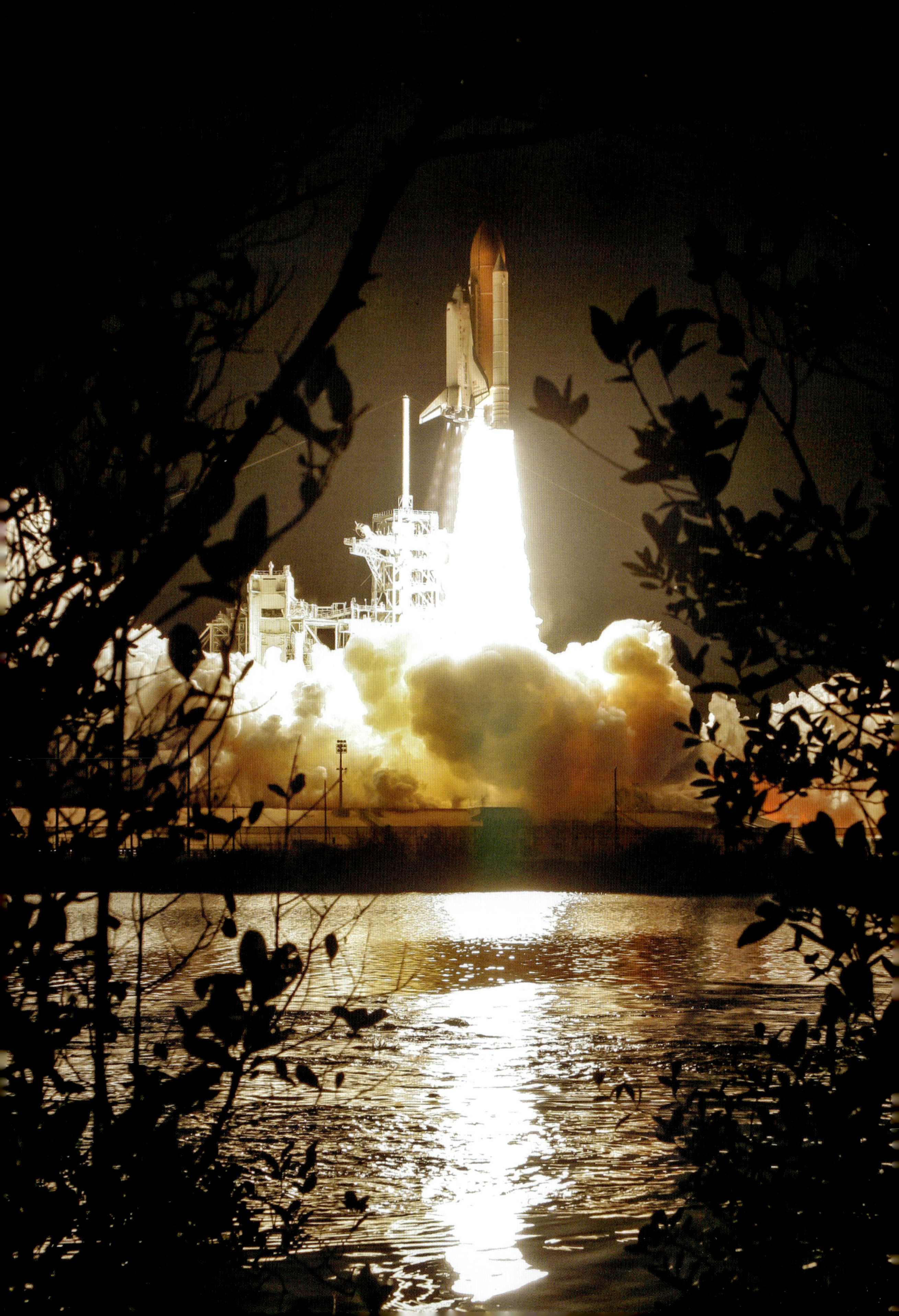

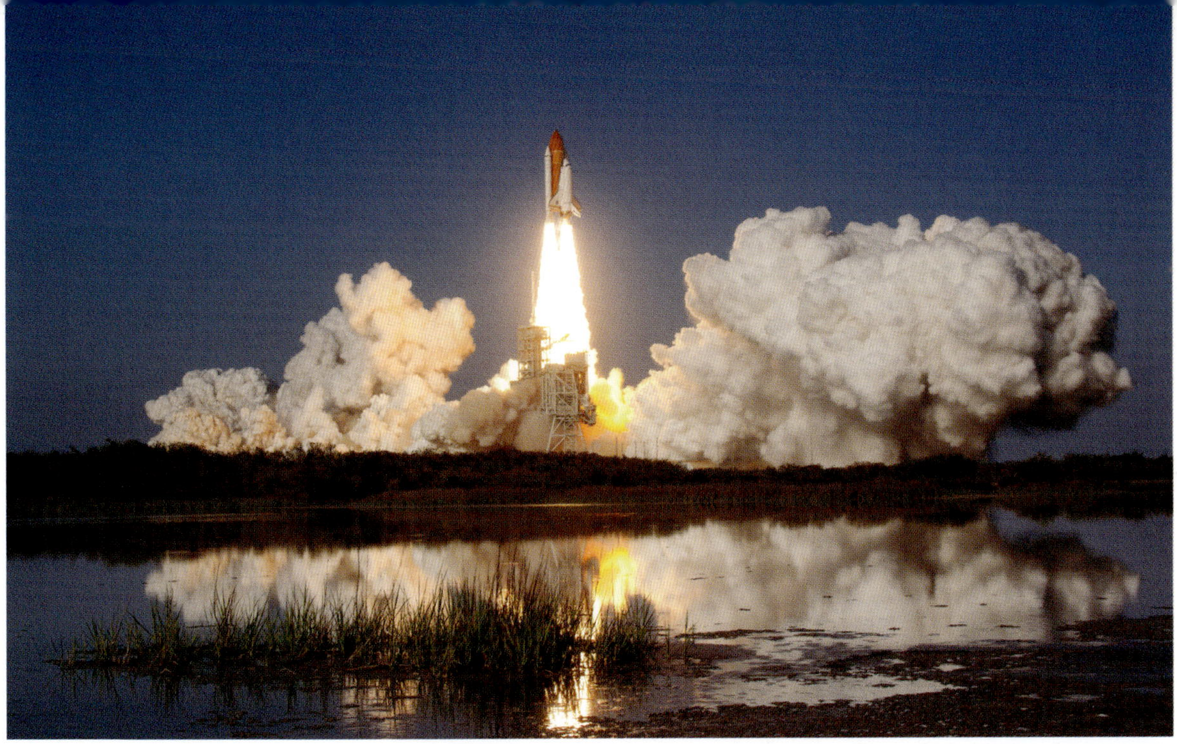

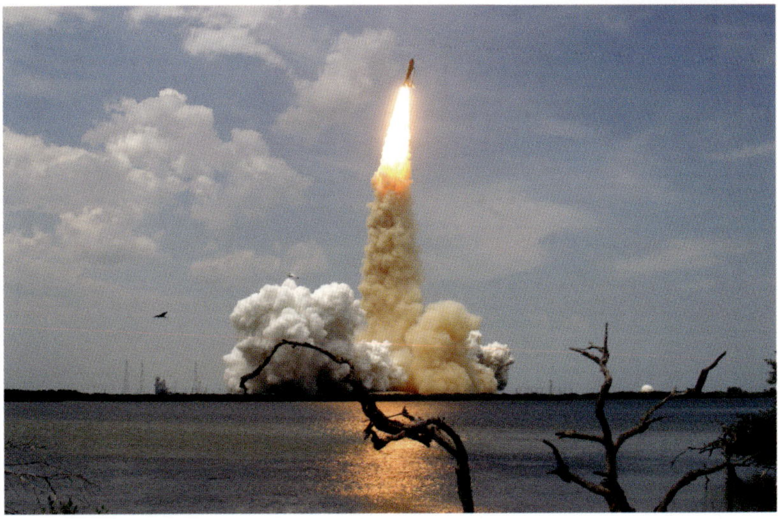

Mangroves

(previous page) Using mangroves to frame the launch photos, both day and night, produced great results. Here, STS-131 Discovery blasts off in 2010.

From the Swamp

(above) This is the type of photo I have in mind when I trudge deep into the soft mud of the surrounding swamp. (See pages 36–37 for a photo taken during the placing of this camera.) This is STS-133 Discovery being reflected in the water on the west side of pad 39A as it begins its last journey into space.

Roll Maneuver

(above) Dead tree branches frame Atlantis as it begins what is known as the roll maneuver shortly after launch. The shuttle rotates so that the orbiter is on the bottom and in line with the trajectory the shuttle will take during its trip into space. For STS-125, seen here, the shuttle made its last service and upgrade mission to the famed Hubble space telescope.

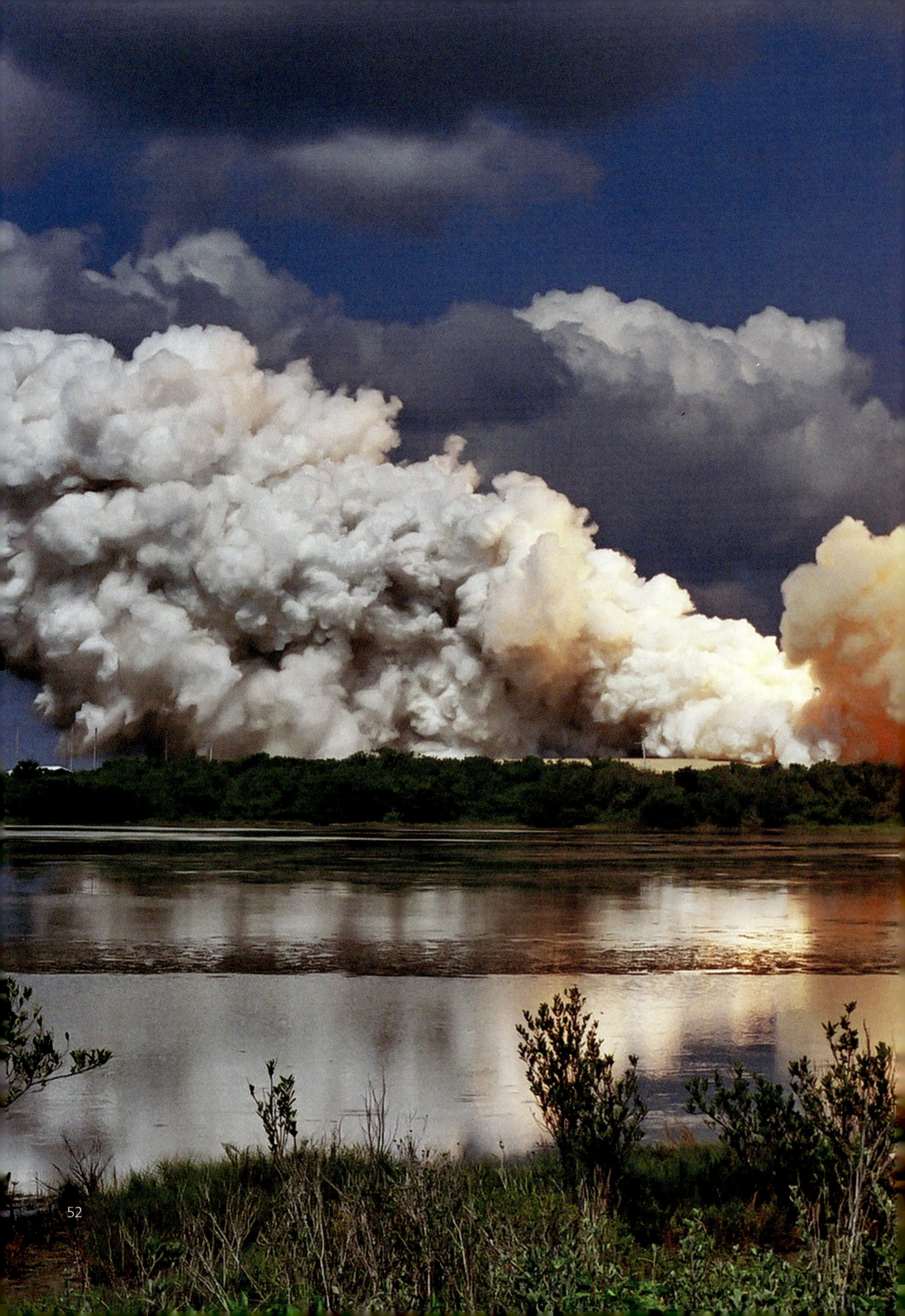

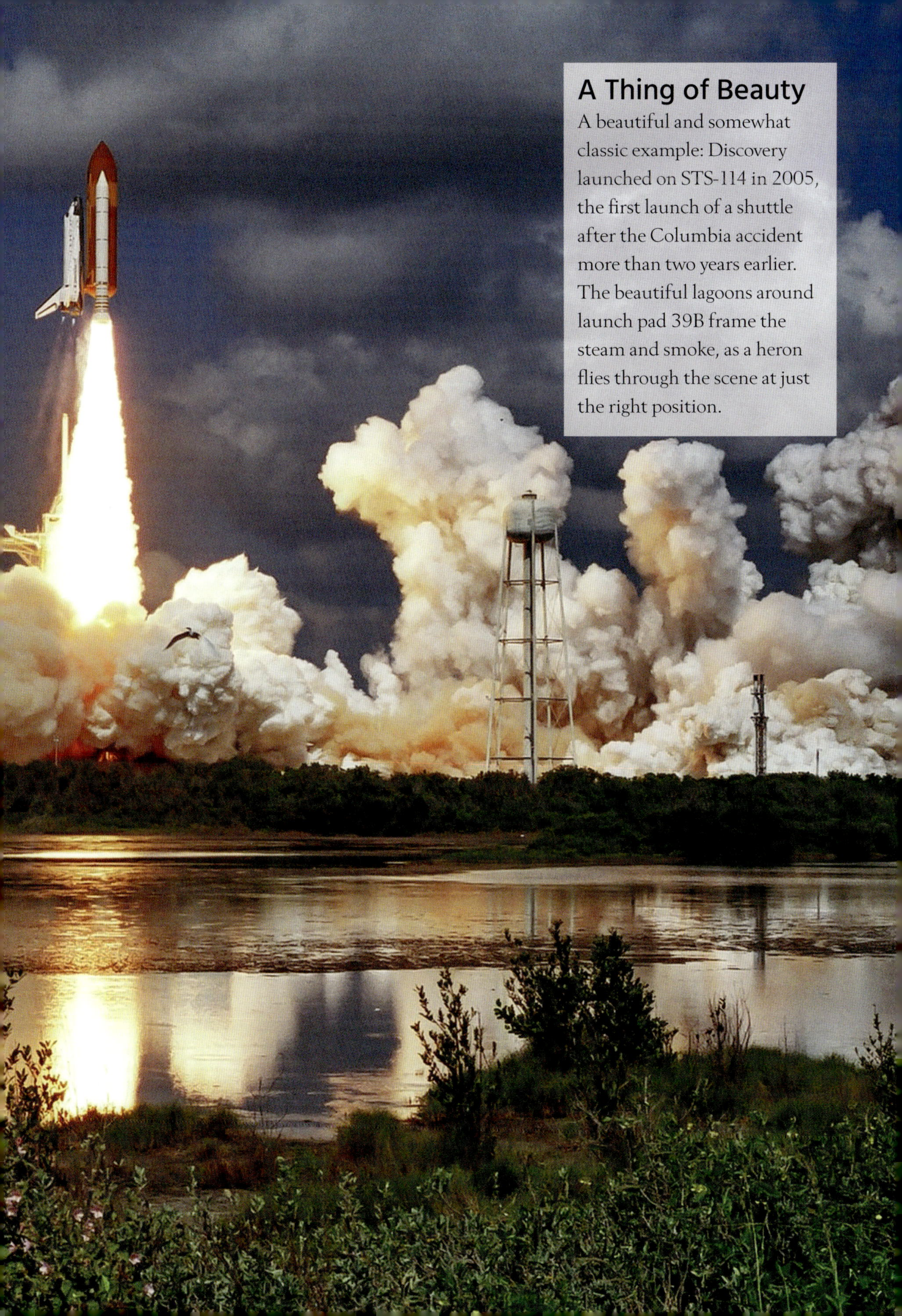

A Thing of Beauty

A beautiful and somewhat classic example: Discovery launched on STS-114 in 2005, the first launch of a shuttle after the Columbia accident more than two years earlier. The beautiful lagoons around launch pad 39B frame the steam and smoke, as a heron flies through the scene at just the right position.

The Arc

(top) The arc, or streak shot, is my favorite way to photograph night launches. At night, many static photos of the liftoff can look the same, as there is no color and no clouds are visible to make them unique. It's hard to remember which launch it was just by looking at a night shot, whereas day launches are more unique as they come with different lighting and clouds.

STS-131 Discovery was one of the most beautiful launches because it took off about 50 minutes before sunrise. This photo was captured at the VIP viewing site.

Before Sunrise

(bottom) The shuttle would rarely take off when it was overcast. This was due to safety reasons: even if the weather is good, thick clouds can produce dangerous static electricity, or hamper the view or signals for the astronauts if they had to return to earth in an emergency. So they would only launch if the clouds were thin enough, and above a certain altitude. The launches of Endeavour on STS-123, seen here, and STS-134, on the following page, were two of those rare launches. The shuttle roared through about as low a cloud deck as it could have safely launched in those days, making the photos unique.

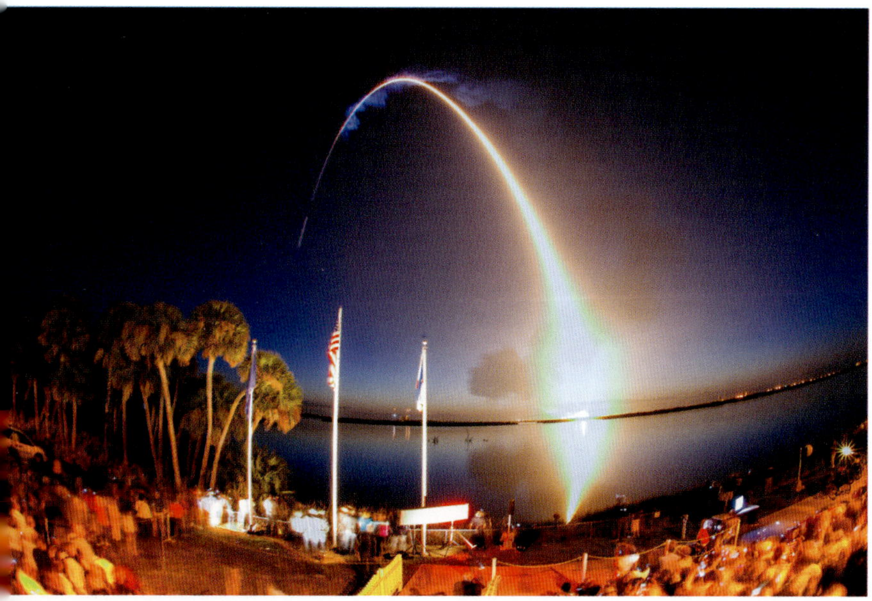

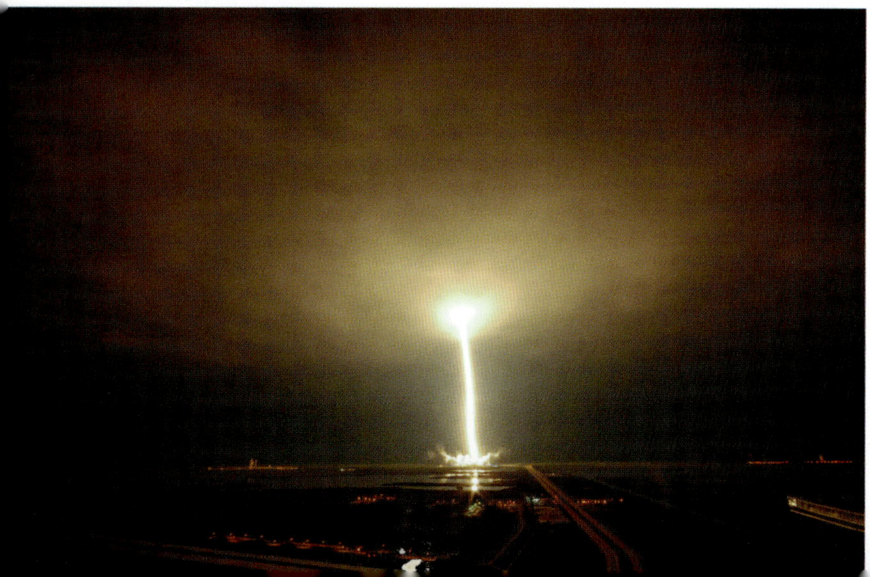

Thin Cloud Cover

(following page) When STS-134 Endeavour took flight, it did so under a low-hanging but thin cloud layer, which is unusual for shuttle launches. After punching through the clouds and out of view of spectators at the press site, the shuttle's smoke trail and its shadow being cast on the clouds from above became the subject of the photo.

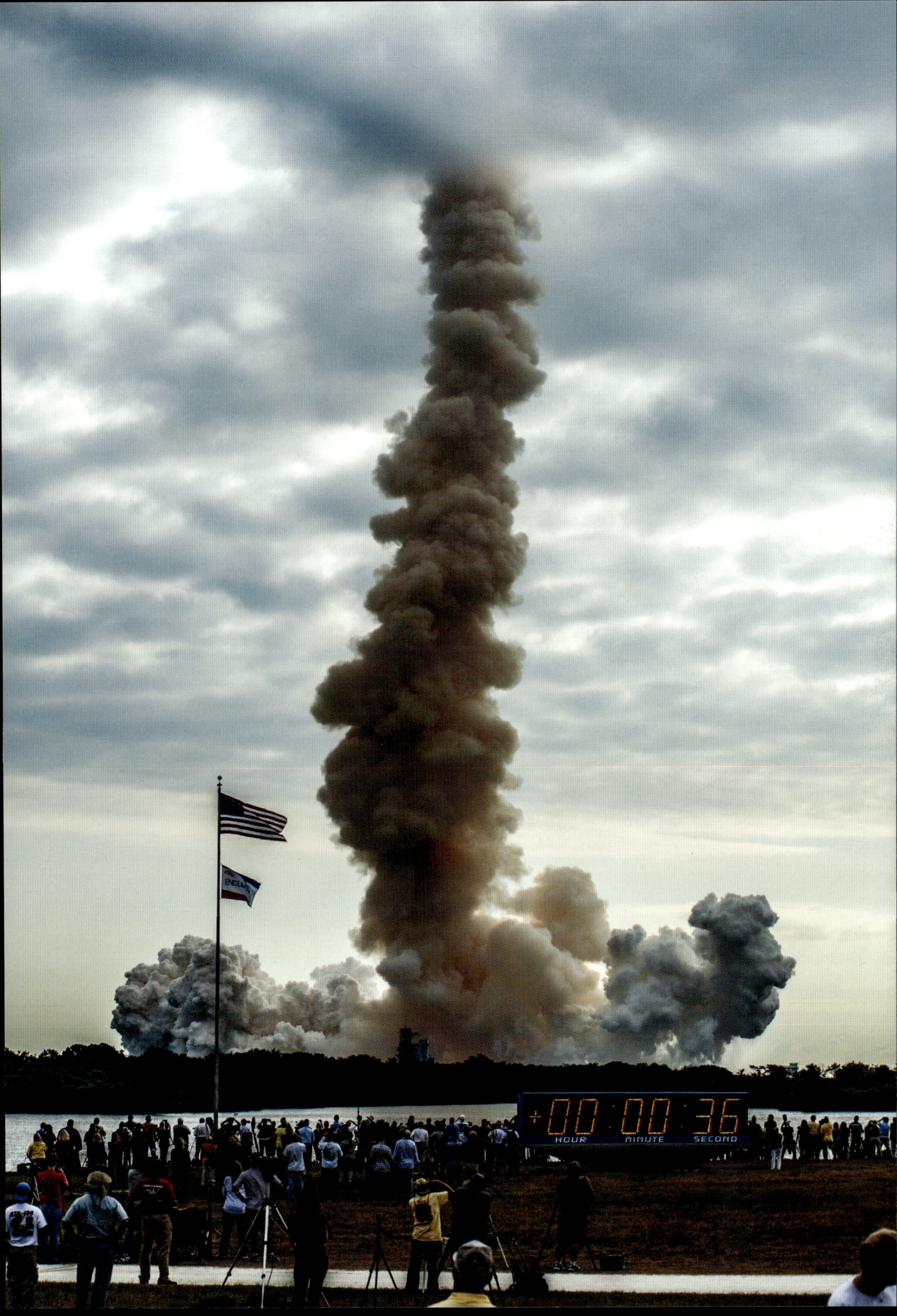
+00 00 36
HOUR MINUTE SECOND

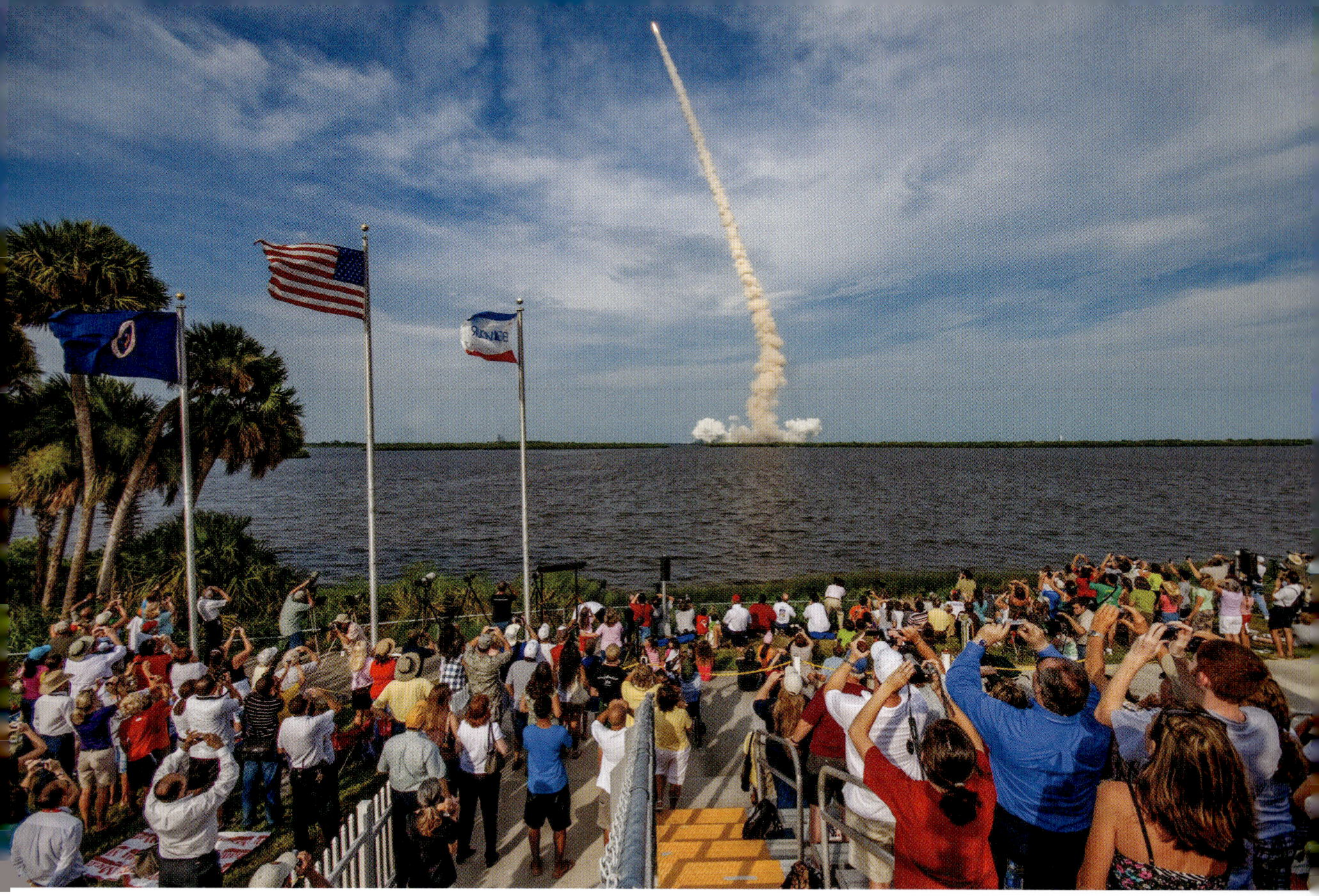
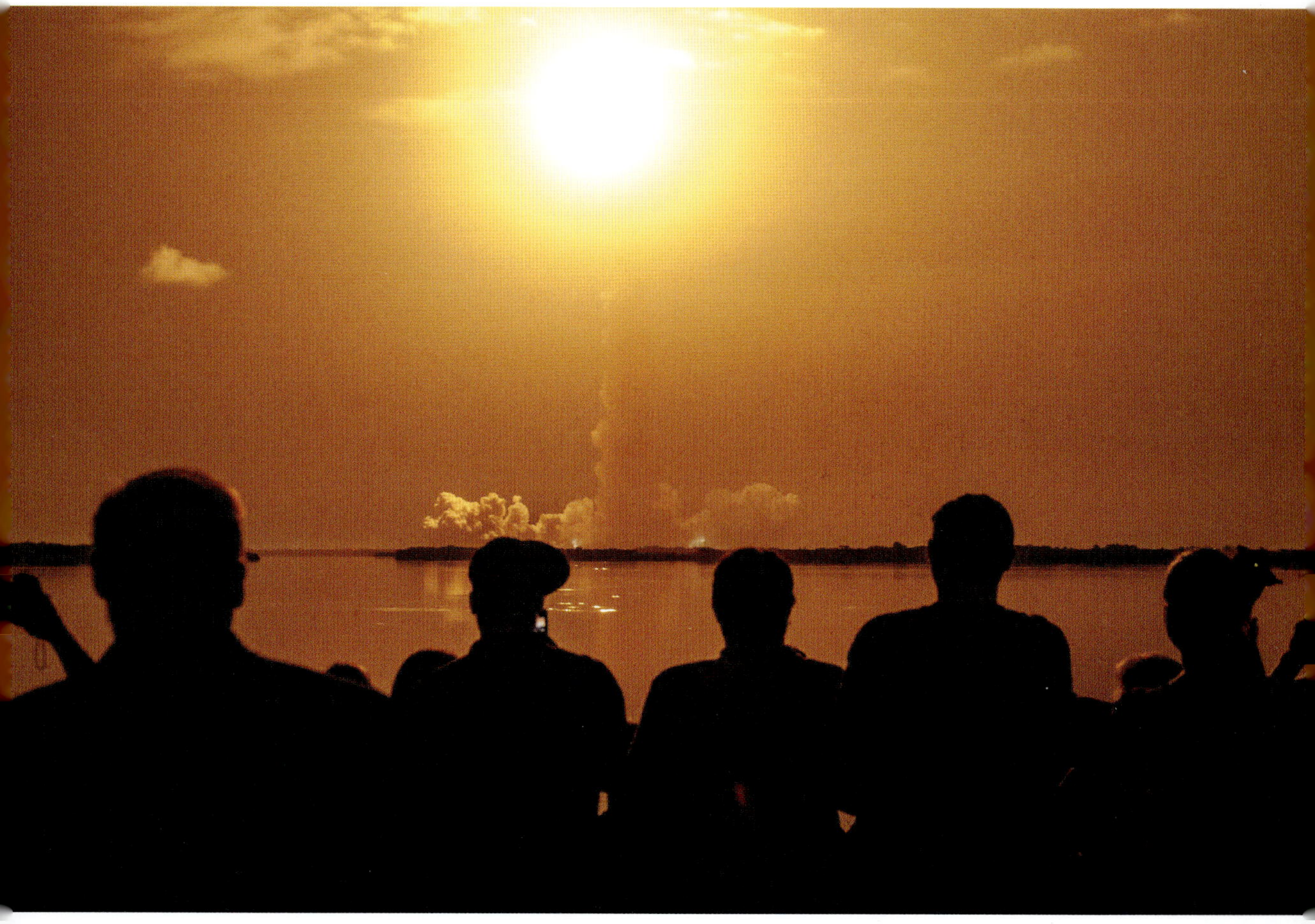

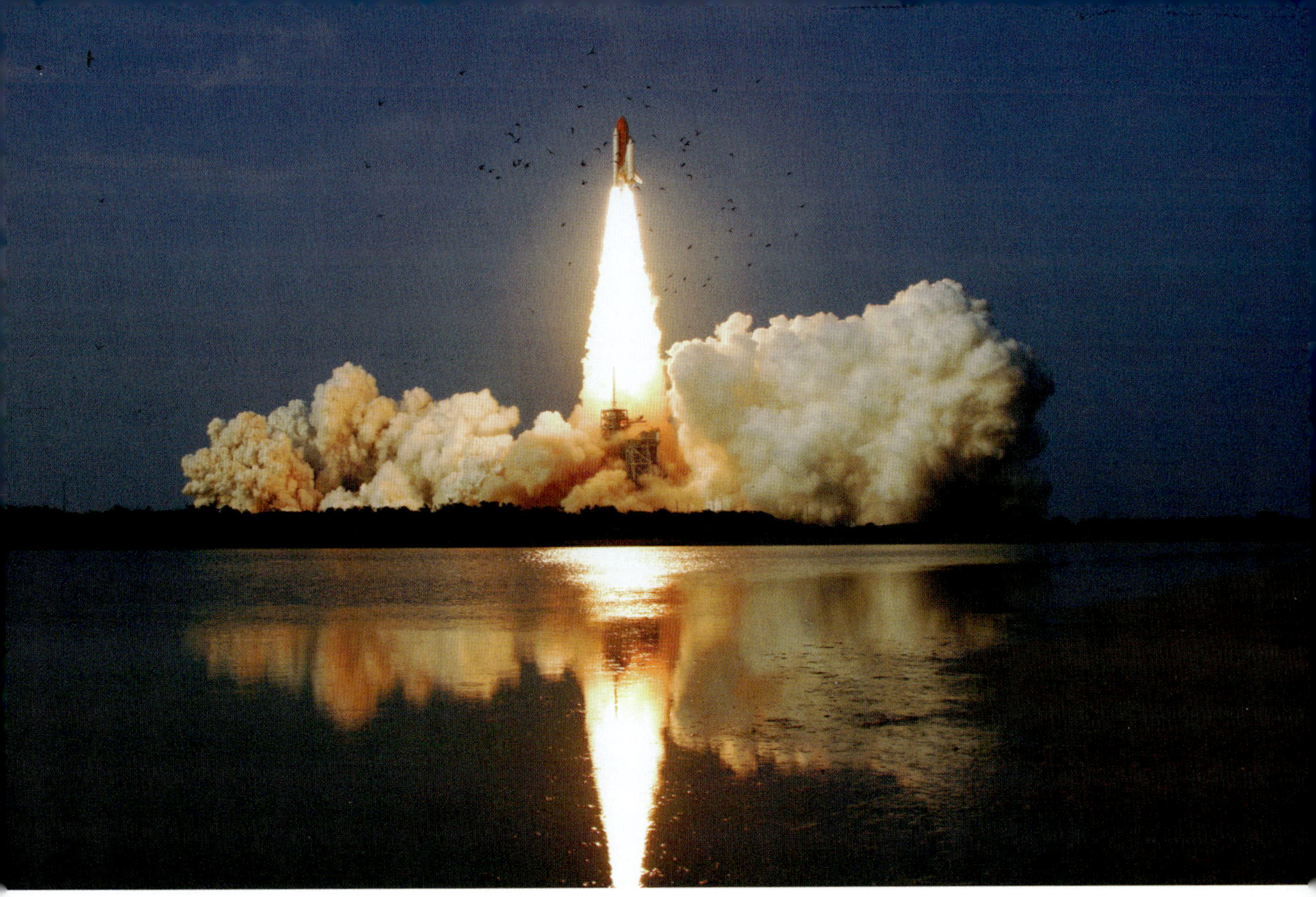

The Crowds

(previous page, top and bottom) The crowds are another element I wanted to capture, both day and night. This sweeping wide-angle shot *(top)* shows thousands at the VIP viewing site looking on at STS-127, Endeavour, as it launched in 2009. At night, this can be more challenging, so I decided to frame a smaller group in silhouette *(bottom)* against the natural light of the launch, which is extremely bright in person.

STS-118 Launch

(above) This is another favorite. As STS-118 launched, this camera, located in a lagoon to the west of the pad, captured a flock of birds flying past the shuttle at just the right moment. The launch took place shortly before sunset and provided great lighting.

The Coastline

(following page spread) The coastline here is part of Canaveral National Seashore, a protected national park system unit. As the shuttle program neared an end, inspired by some earlier program photos, I wanted to capture one launch from this perspective, which is difficult, as there are not many areas where you can get a clean shot of both the pad and the ocean. Here, STS-134 Endeavour arcs over the Atlantic on its way into orbit.

The launch took place shortly before sunset and provided great lighting.

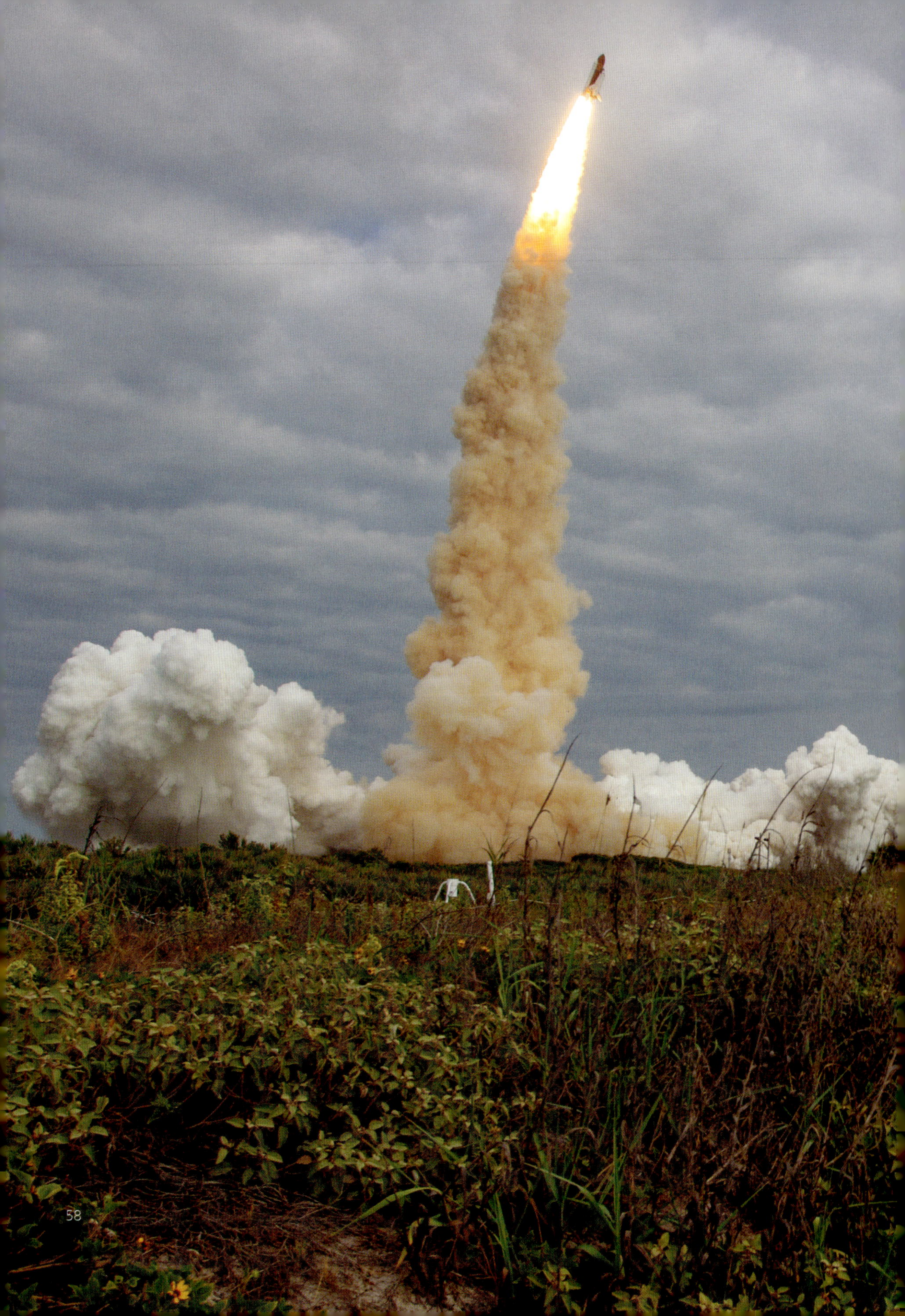

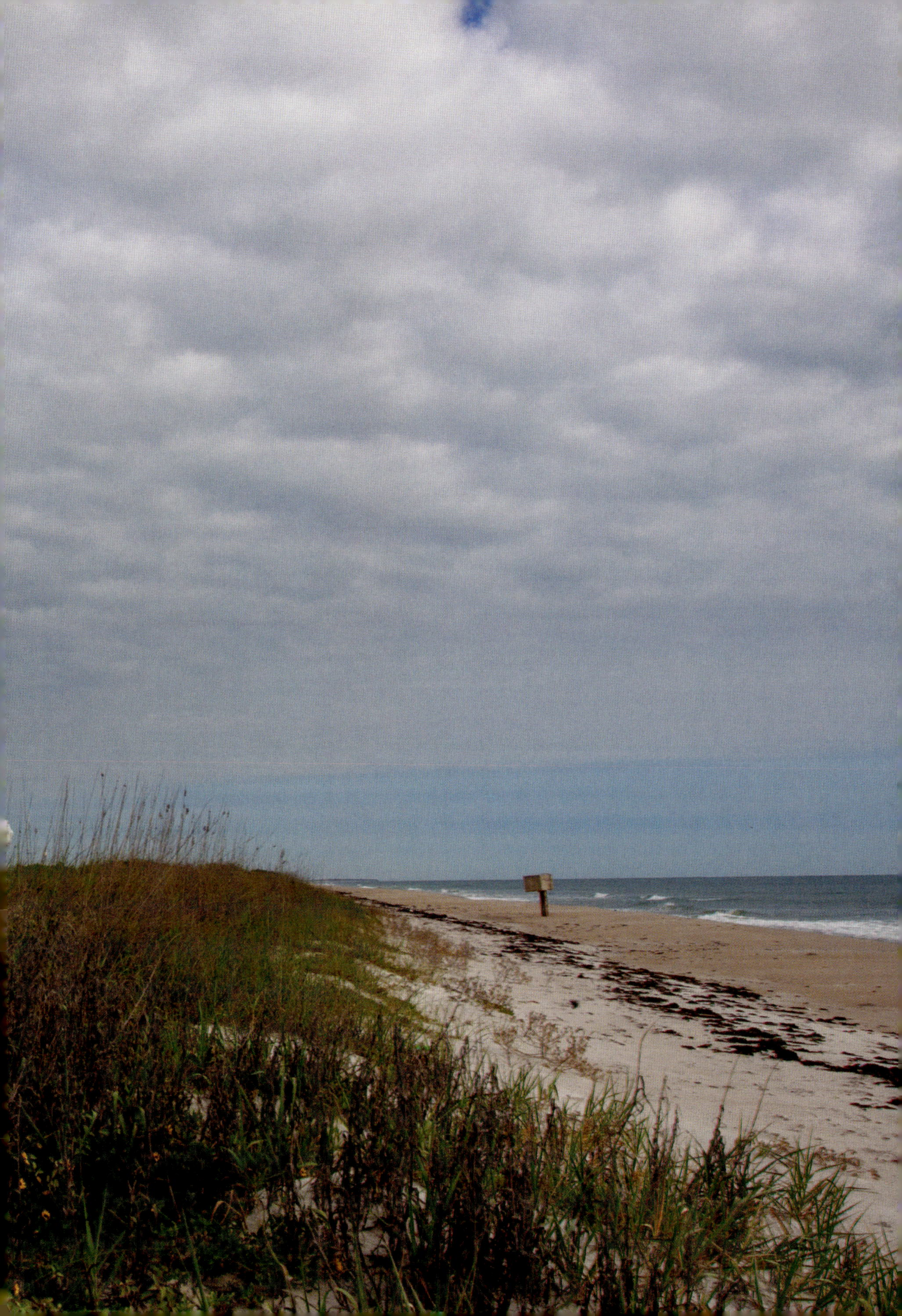

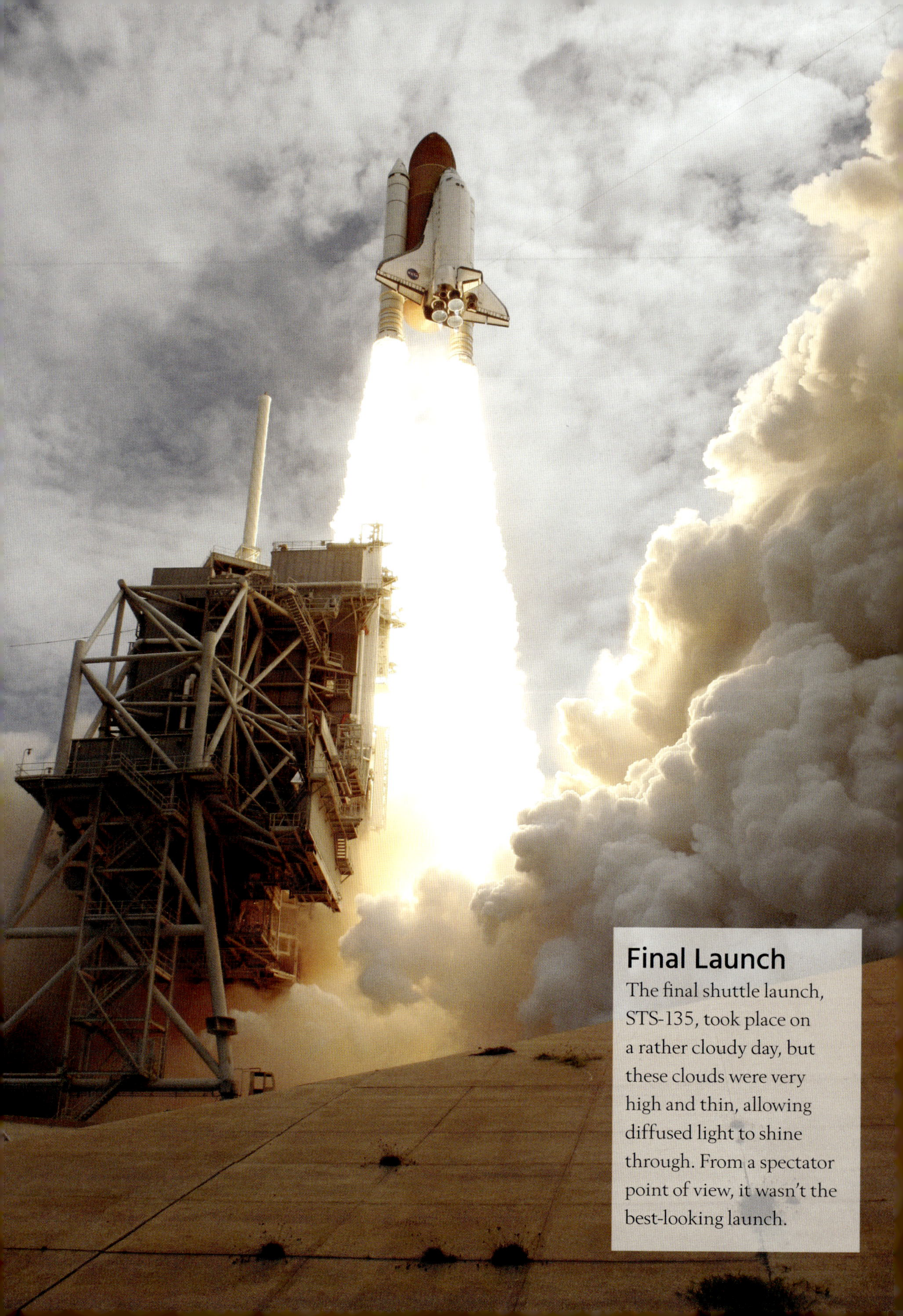

Final Launch

The final shuttle launch, STS-135, took place on a rather cloudy day, but these clouds were very high and thin, allowing diffused light to shine through. From a spectator point of view, it wasn't the best-looking launch.

Roll Program

Atlantis is in the middle of its roll-program maneuver, when the orbiter goes from facing south on the ground to facing the direction of travel, upside-down, on its way into space. At this moment, the camera caught the tail pointed precisely toward it as smoke billowed in all directions.

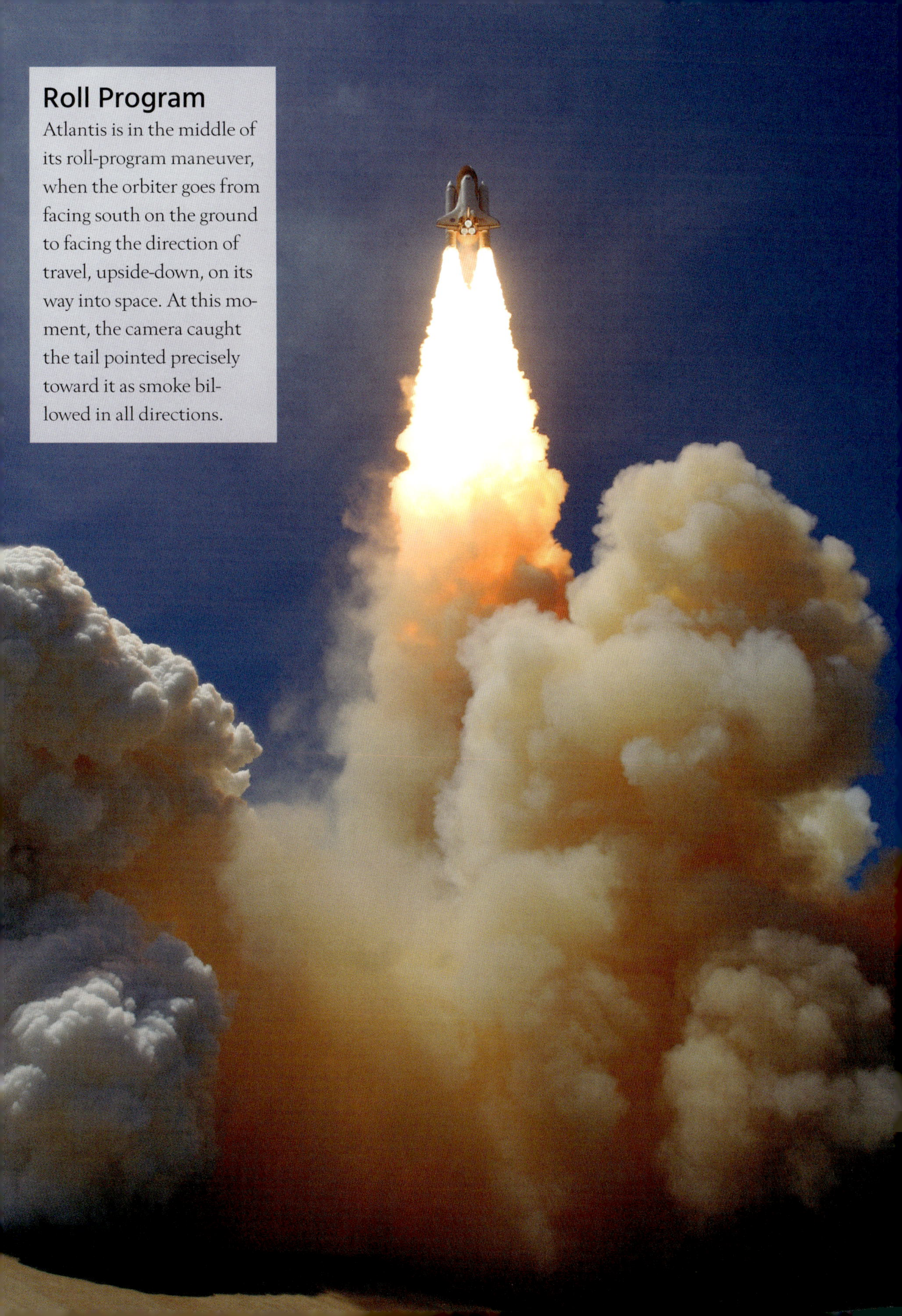

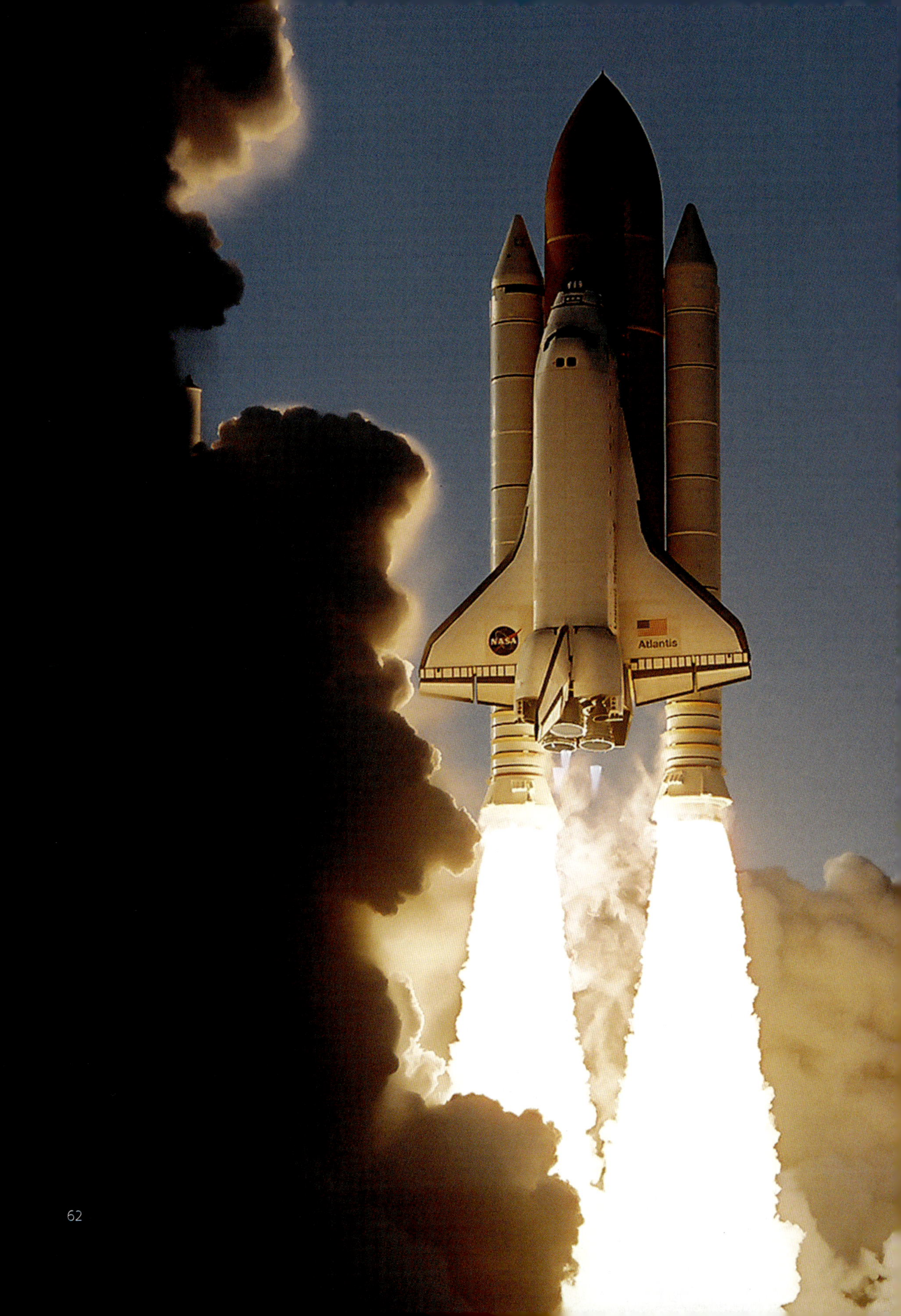
NASA
Atlantis

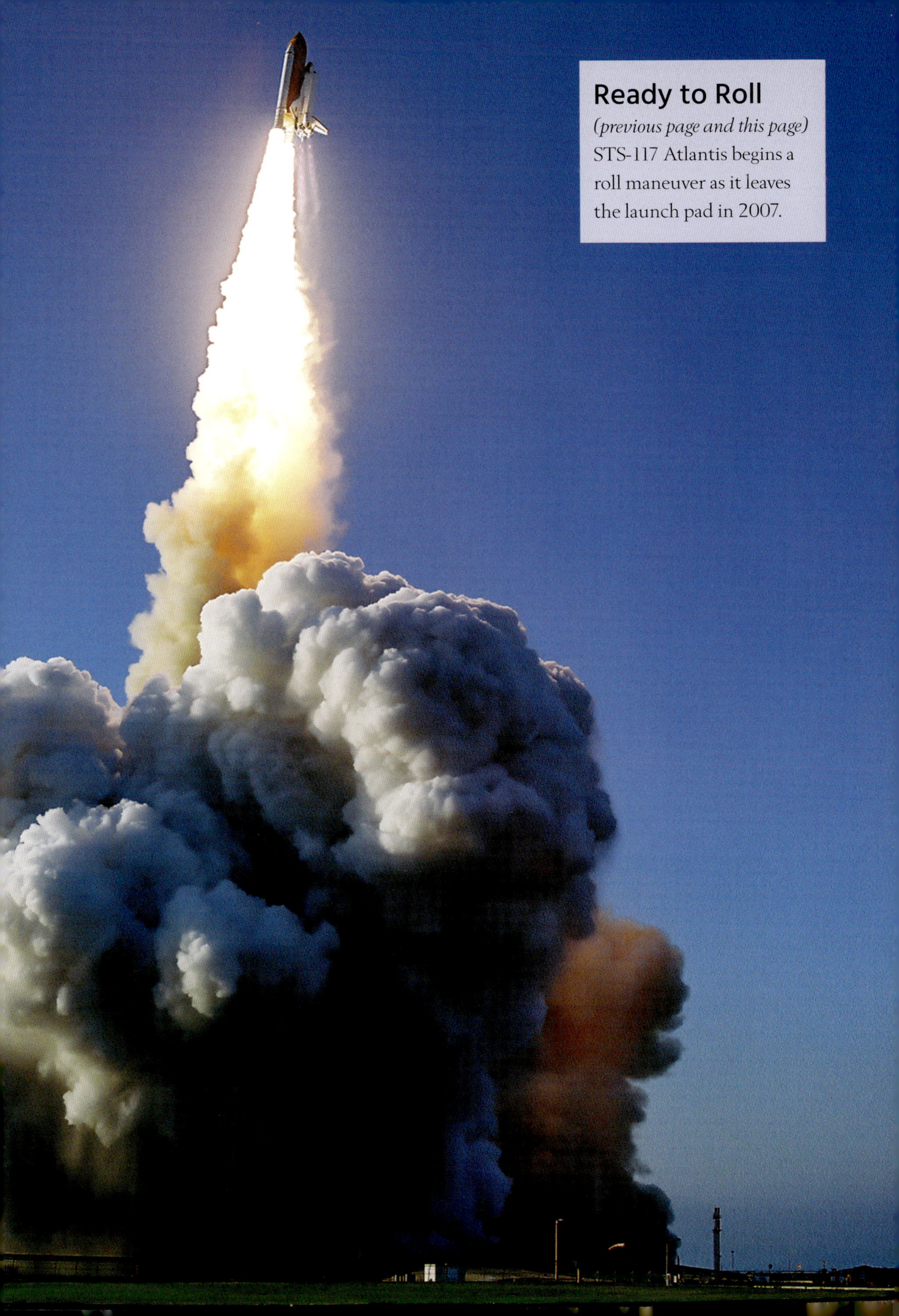

Ready to Roll

(previous page and this page) STS-117 Atlantis begins a roll maneuver as it leaves the launch pad in 2007.

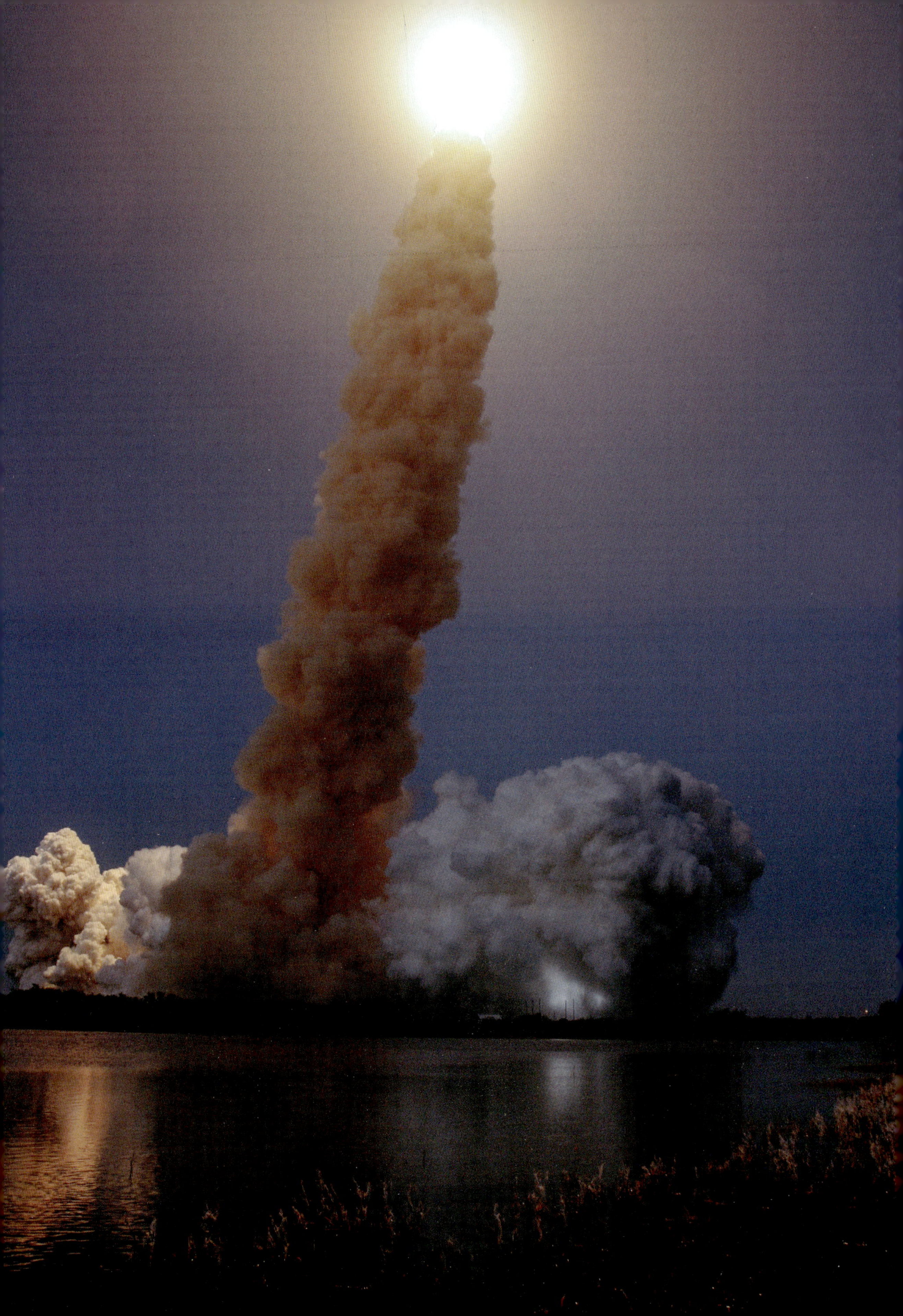

Twilight Launch

(previous page) During a twilight launch, I was challenged to capture the light in the sky. At proper exposures for the shuttle, it looked like a night launch with a black sky. There is a point at which exposing for the flames negates what the eye sees during twilight launches, and the photos are less memorable. So whenever we had one of these uncommon launches, I did my best to find a way to capture the beauty and light, even if it meant the shuttle itself would be less visible. What we see here is more natural (the launch flames are *really* bright to the naked eye), as STS-119 Discovery leaves its swamp-surrounded launch pad.

Night Launch

(top) In this photo of STS-126 Endeavour, I wanted to capture what a night launch looks like to the naked eye. Exposing for this is best done when there are high clouds, as they get lit up and form a nice backdrop with some depth to the photo. I cut out the horizon here except for this viewing platform at the shuttle landing runway with a few people on it, all looking up in wonder.

(bottom) Discovery lifts off on STS-131, as seen from across the water.

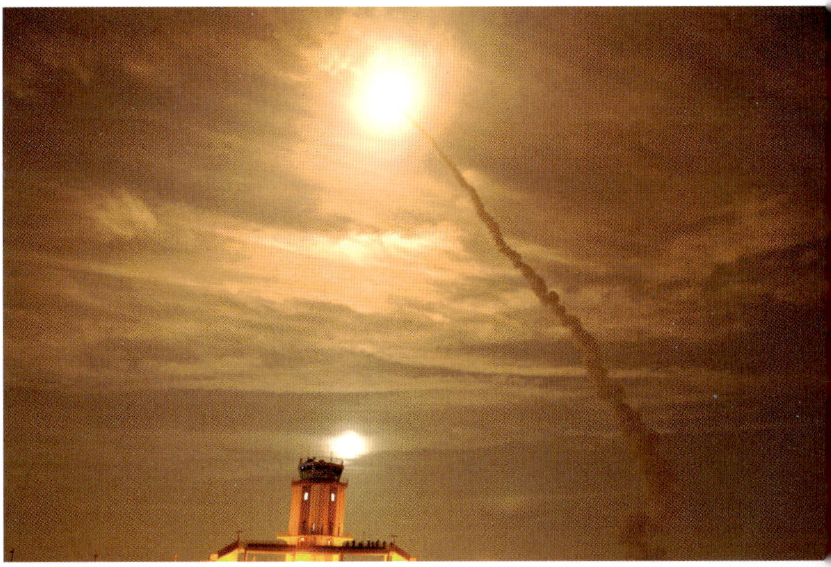

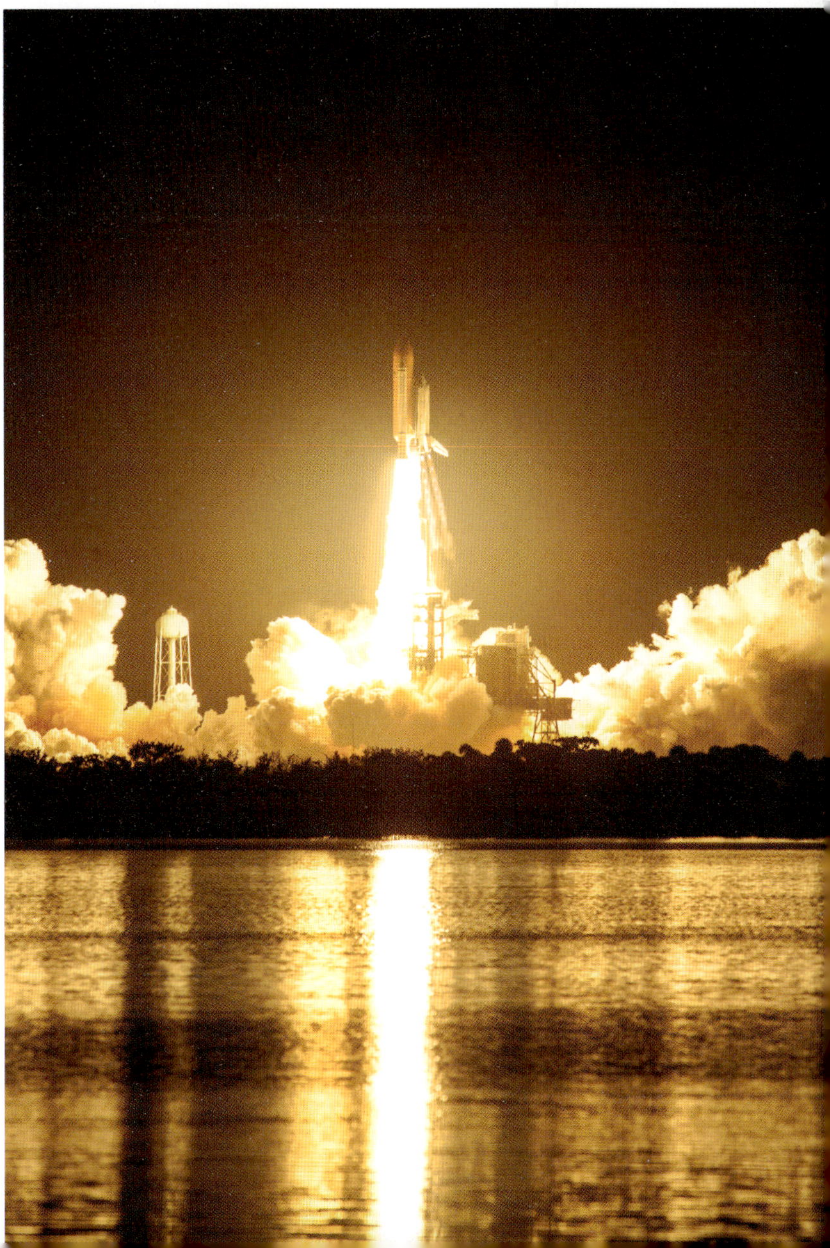

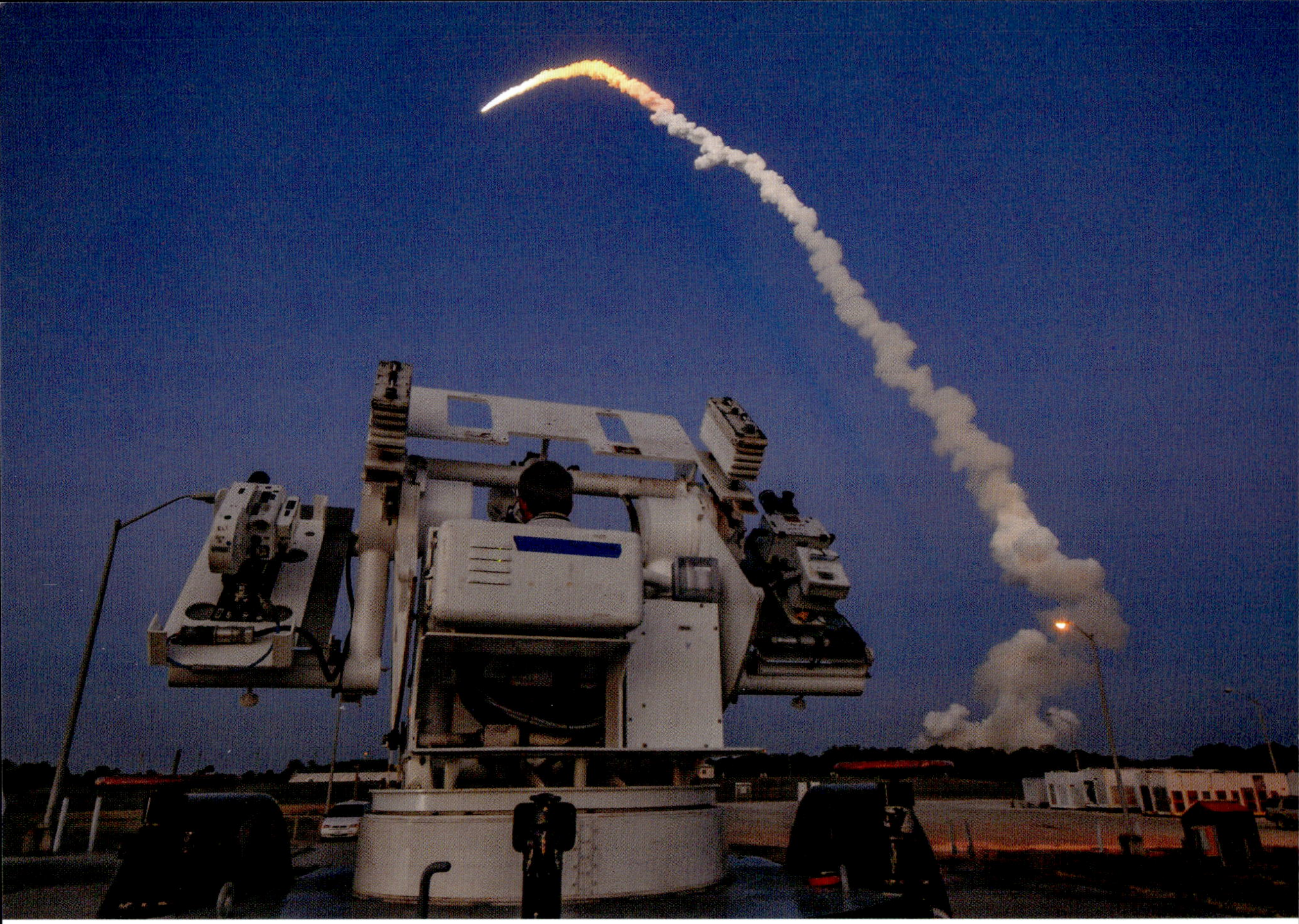

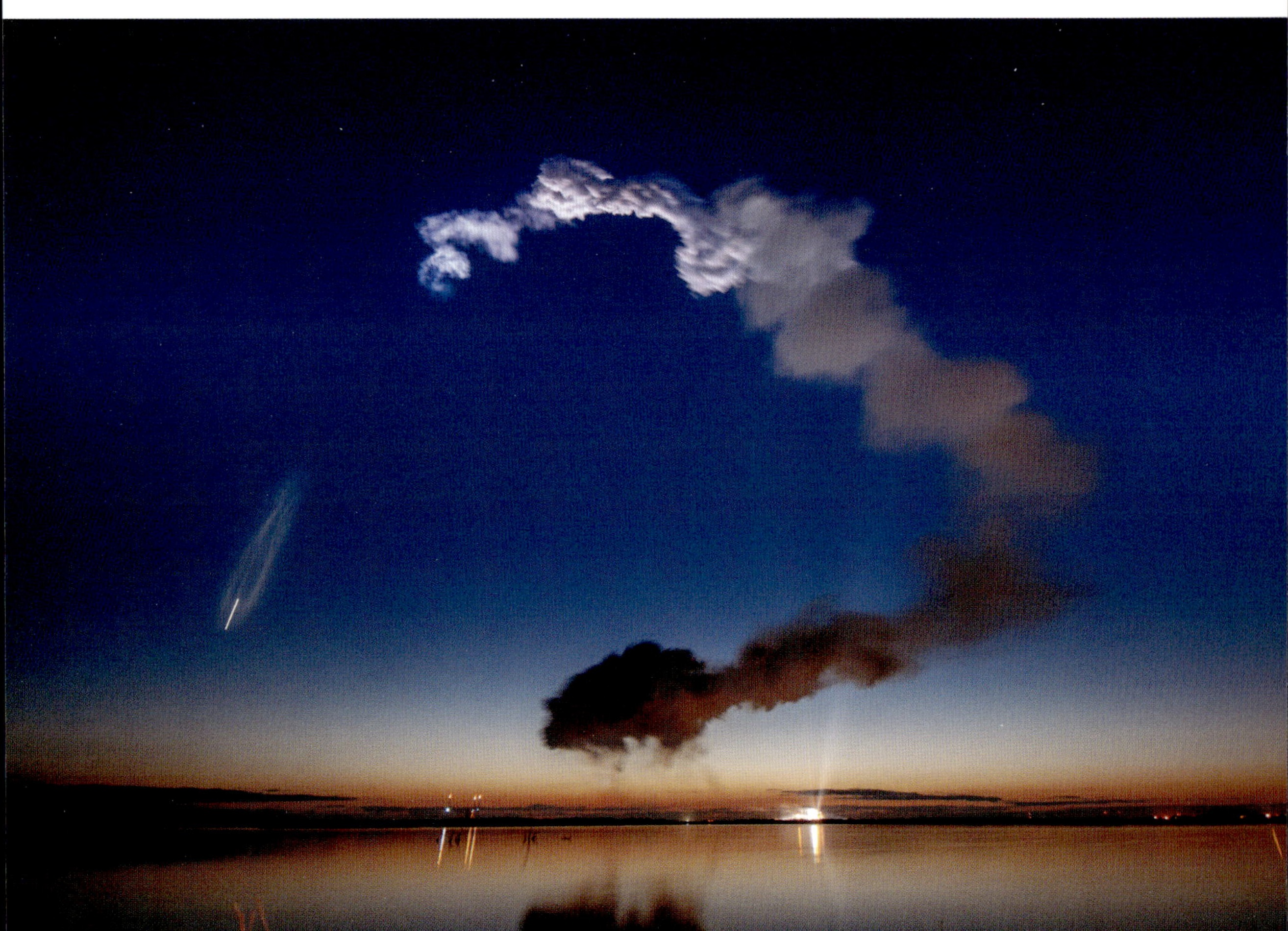

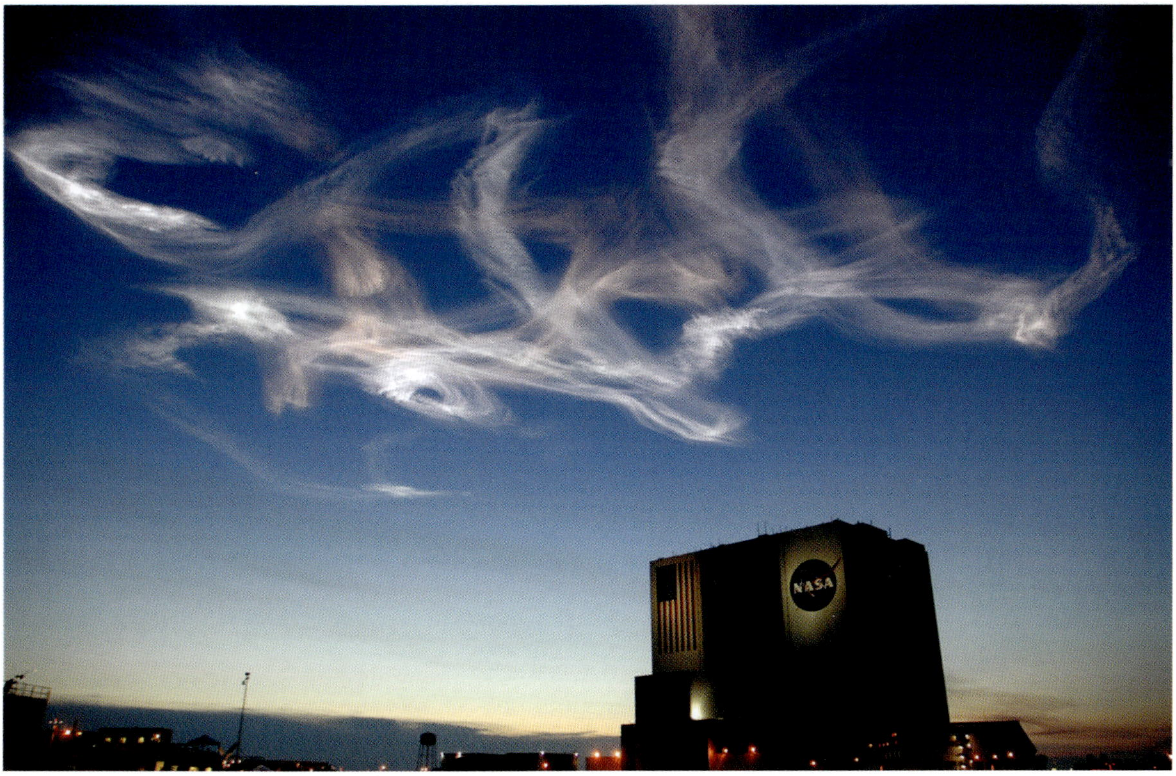

A Grand Spectacle

(previous page, top) Launches that take place during twilight, before sunrise or after sunset, are the rarest and the most beautiful. As the shuttle lifts off in darkness, it rises and eventually hits the sunlight, its smoke trail turning a rainbow of colors as it transitions from sunset colors into full white. It's the same reason we see it on the ground, as the sun gets lower and turns more orange or red in color due to the thicker atmosphere we are observing it through. Here, Discovery launches on STS-119 as one of NASA's many manned tracking cameras follows along.

(previous page, bottom) For STS-131, Discovery took off nearly 50 minutes before sunrise. As it headed for the horizon, ice crystals forming from its water-vapor exhaust reflected the sunlight, allowing viewers to see the shuttle as long as possible. The short streak is the glow of the shuttle's main engines over the course of several seconds. At this stage in the launch, it is about 600 miles northeast of the launch site.

Exhaust

(above) One spectacular effect of twilight launches can be the way the exhaust behaves well after the launch is over. Particularly memorable was STS-117 Atlantis. Upper-level winds high in the atmosphere blew its trail in various directions; from the launch site, looking up into the formation, it appeared almost as a spiral or flower in shape. From far away, however, it would appear to be a more vertical display, with lines zig-zagging from top to bottom in altitude.

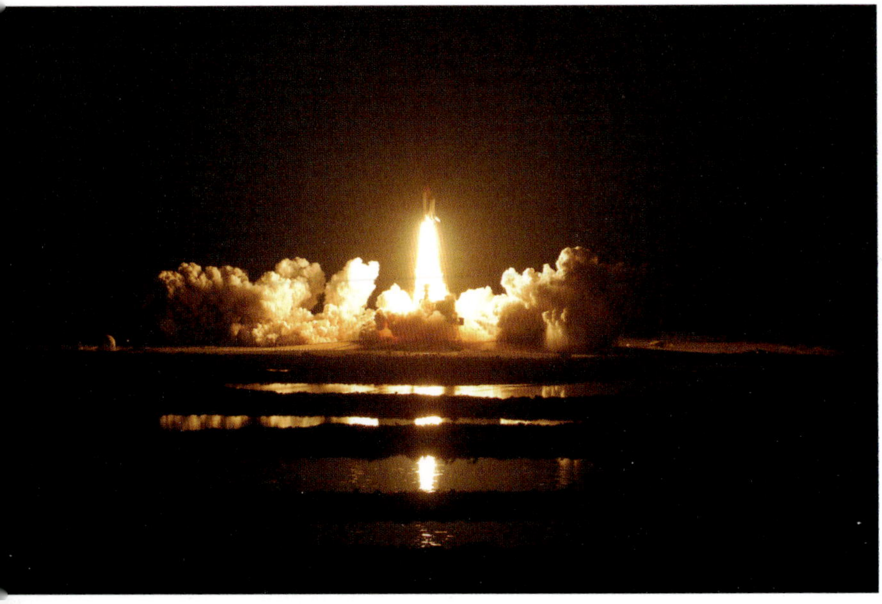

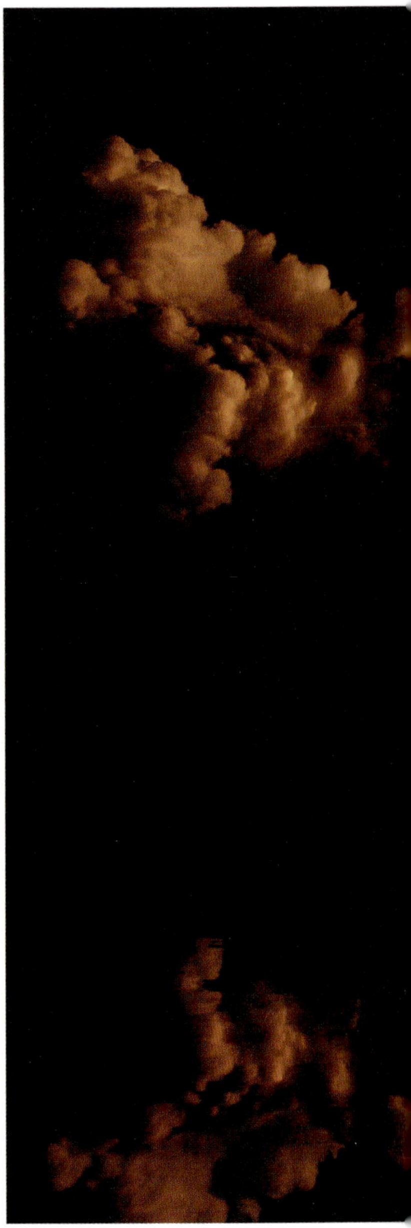

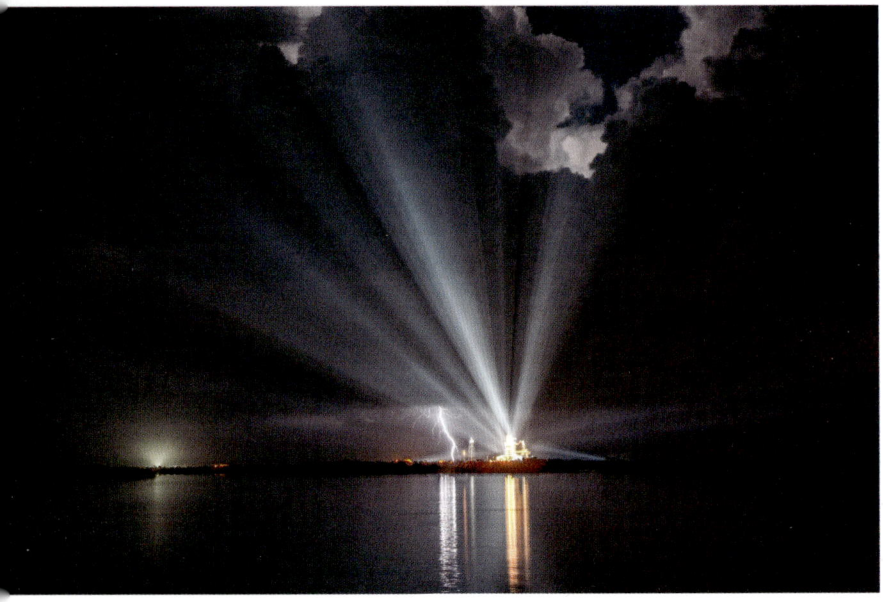

Dramatic Shots

(top left) Night launches are not hard to photograph; there's a known exposure that works great and may differ from rocket to shuttle. But in general, finding a wide variety of great shots is more limited only because of the foreground. For a reflection and a slightly elevated angle, I shot the launch of STS-123 Endeavour with a handheld camera from the roof of the VAB three miles away. With no wind that night, the water was perfectly still before the pad.

(bottom left) Sometimes, you can get a dramatic shot when shuttles don't launch as well. Florida is a hot zone for thunderstorm activity, especially in the summer. Storms frequently form in the afternoon and make their way out to sea as the evening progresses; other storms

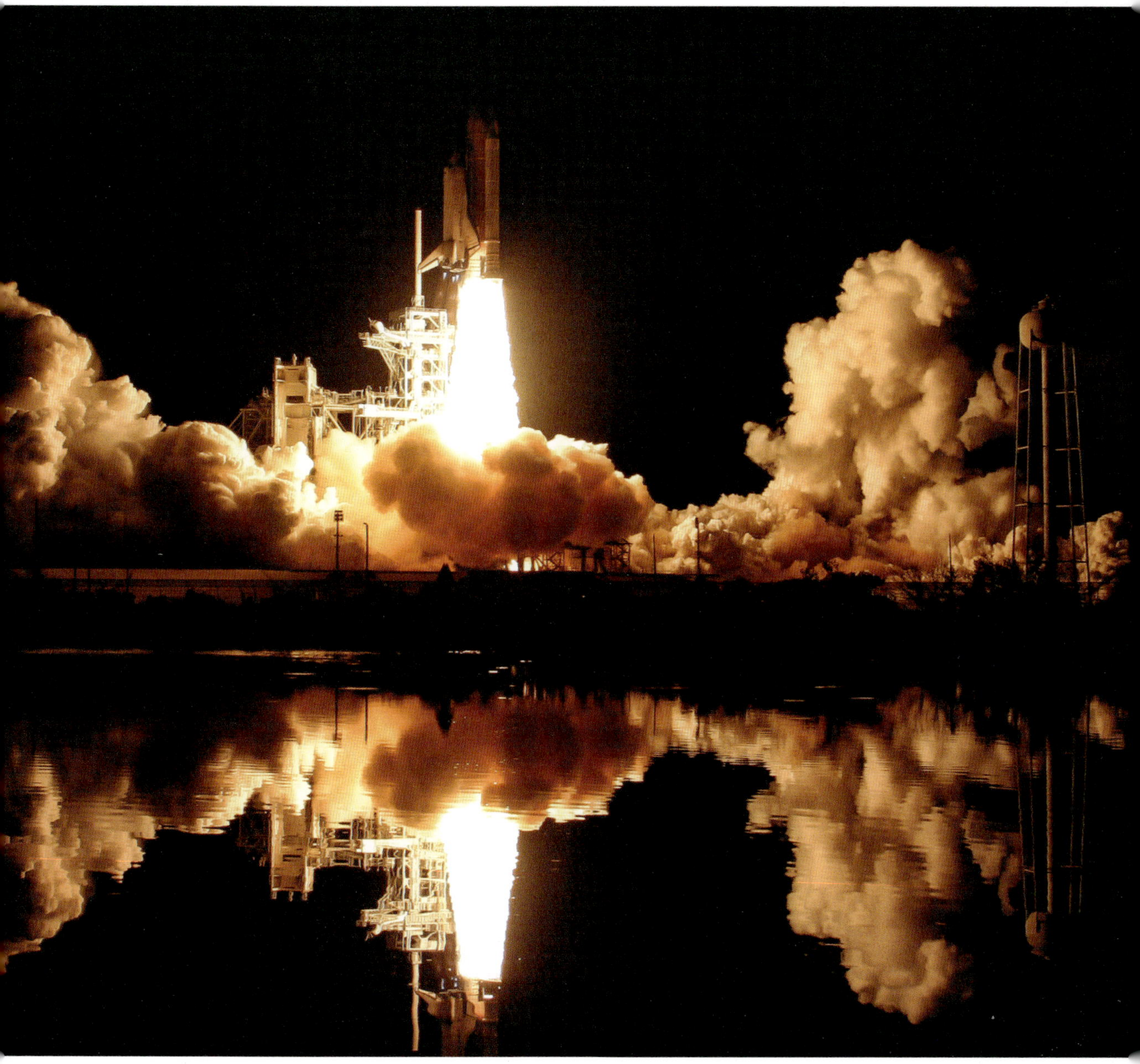

form at night over the gulf stream out in the Atlantic Ocean. On a launch attempt for STS-128, which was scrubbed for these reasons, a bolt of lighting appears to strike the pad (it's actually many miles behind it), while the powerful xenon spotlights on the pad beam their way into the sky.

Mirror Image

(above) A complete lack of air movement in the middle of the night made mirror images of Endeavour as it launched on STS-123 in March 2008. It's rare to have the water this perfect.

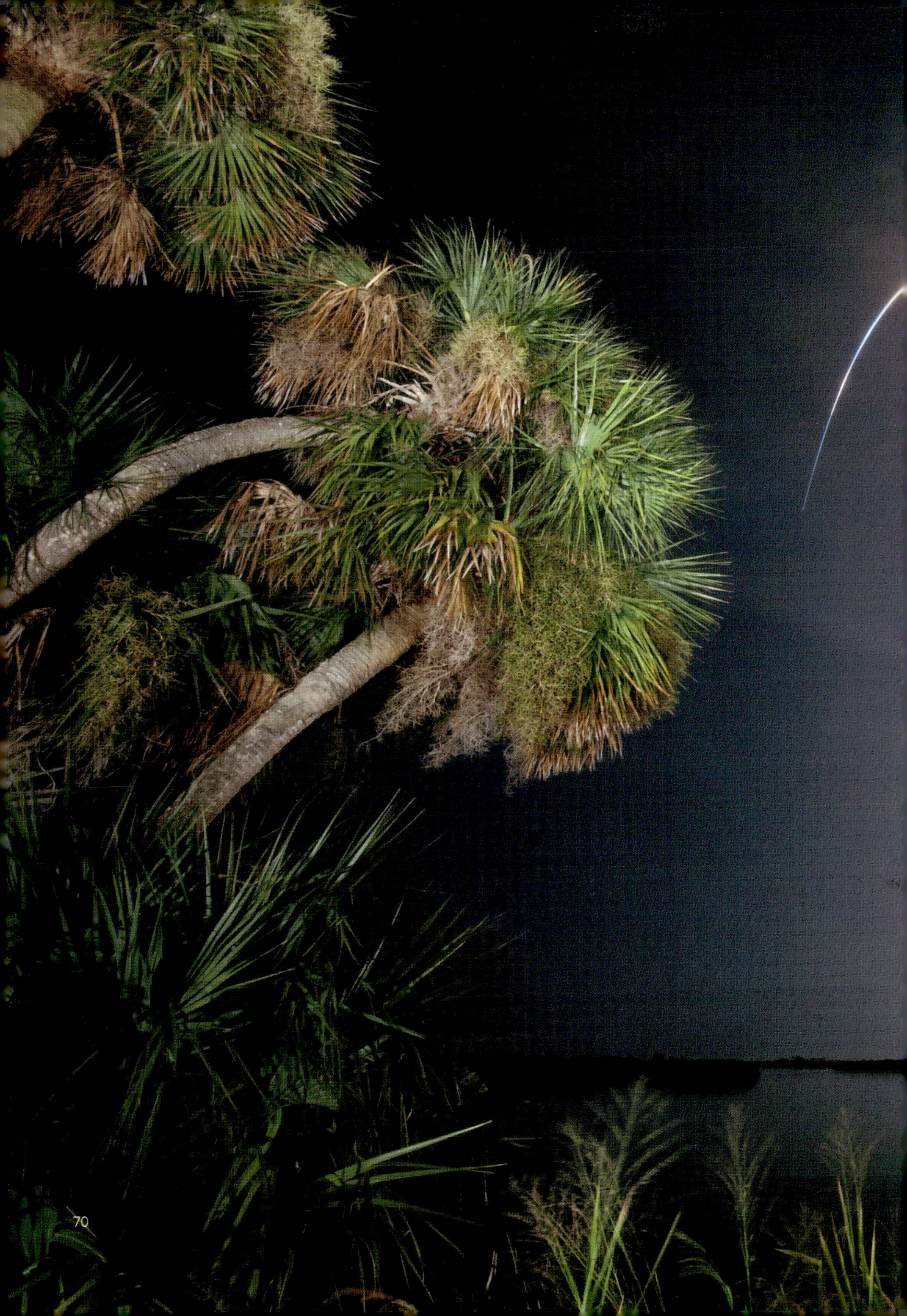

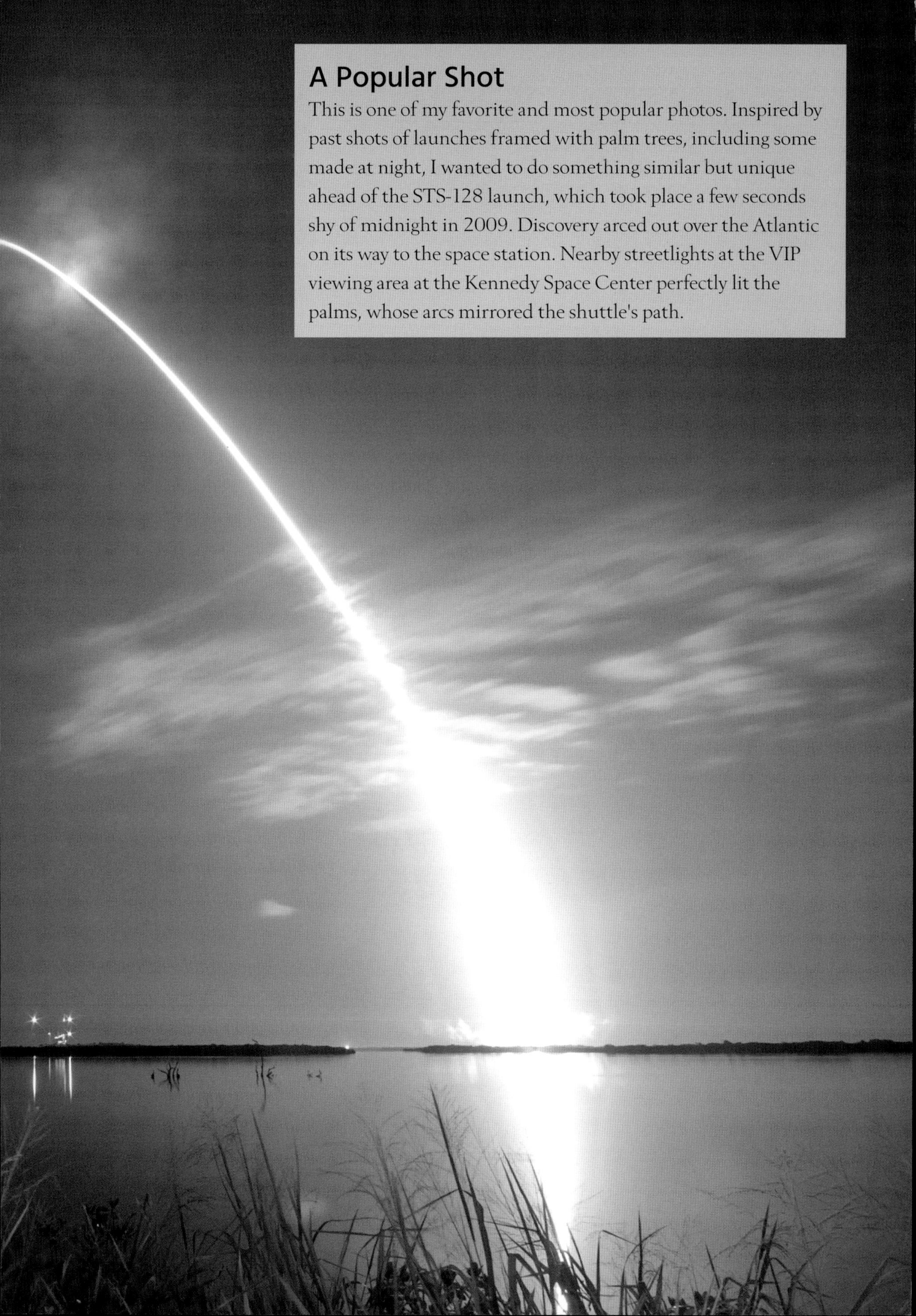

A Popular Shot

This is one of my favorite and most popular photos. Inspired by past shots of launches framed with palm trees, including some made at night, I wanted to do something similar but unique ahead of the STS-128 launch, which took place a few seconds shy of midnight in 2009. Discovery arced out over the Atlantic on its way to the space station. Nearby streetlights at the VIP viewing area at the Kennedy Space Center perfectly lit the palms, whose arcs mirrored the shuttle's path.

RETURN & RETIREMENT

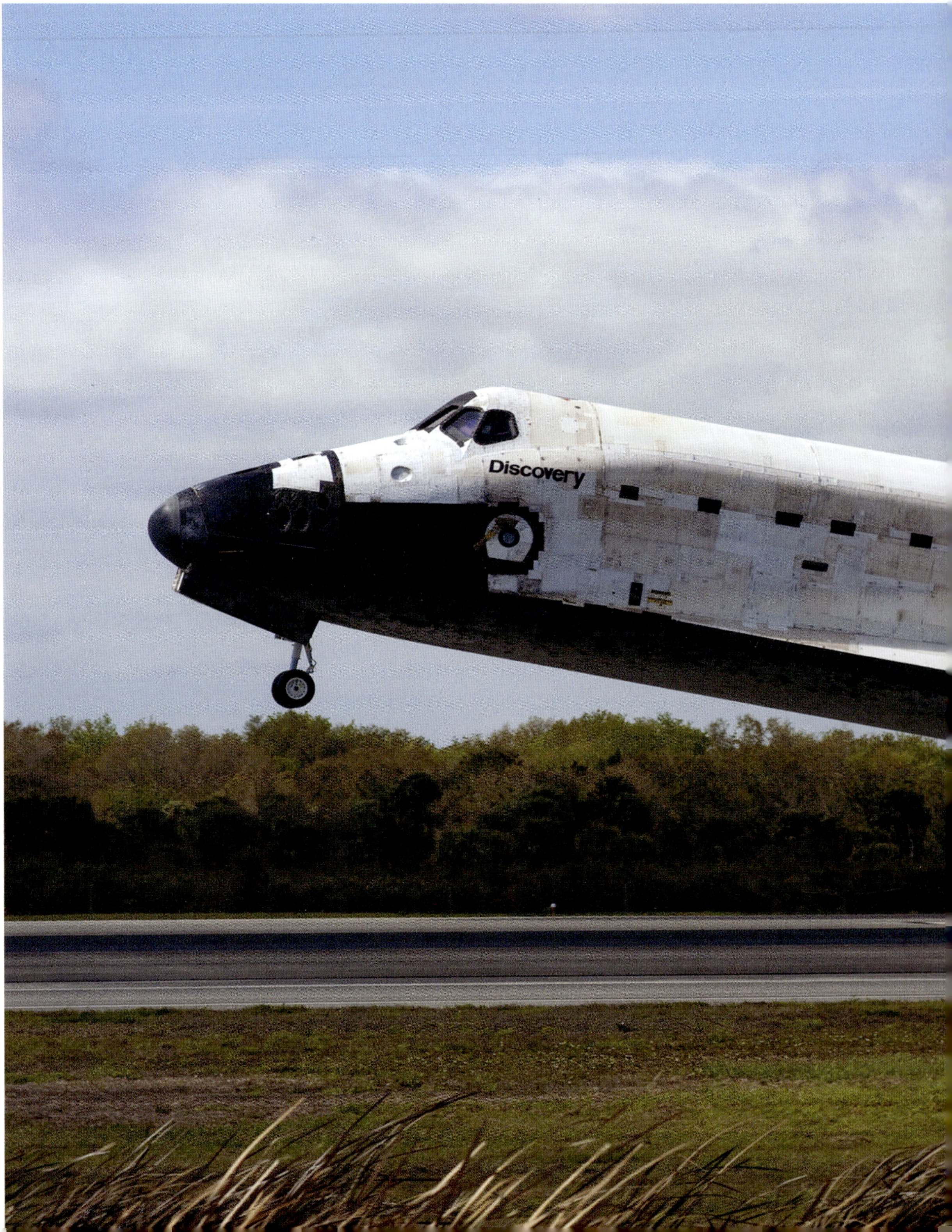

Retirement Age

(below) A shuttle's life comes full cycle when it returns from a mission, which can be anywhere from a few days to more than two weeks after launch. Here, Discovery lands at the Kennedy Space Center's SLF after a visit to the space station on STS-133 in 2011. In this case, it was the orbiter's final mission before retirement.

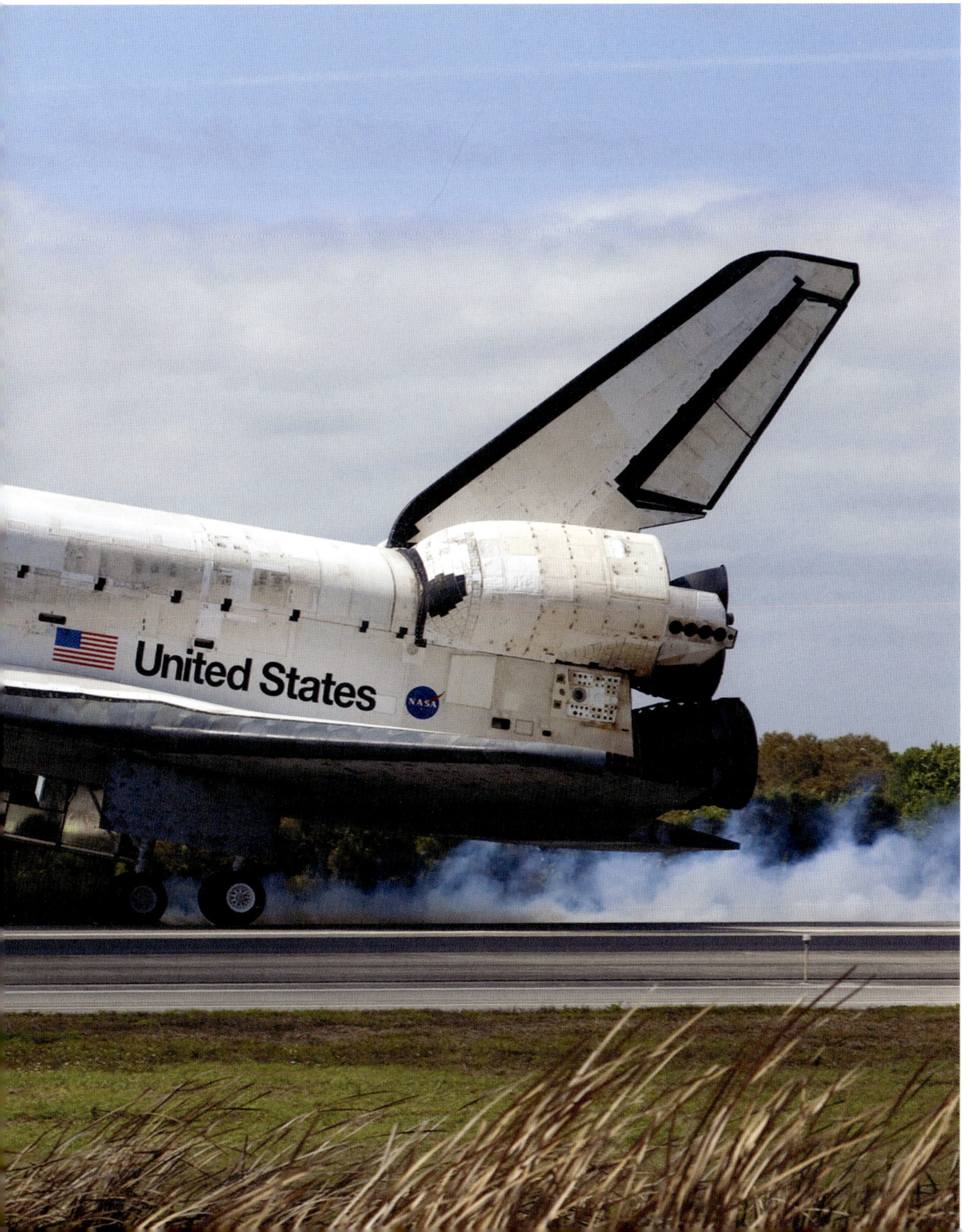

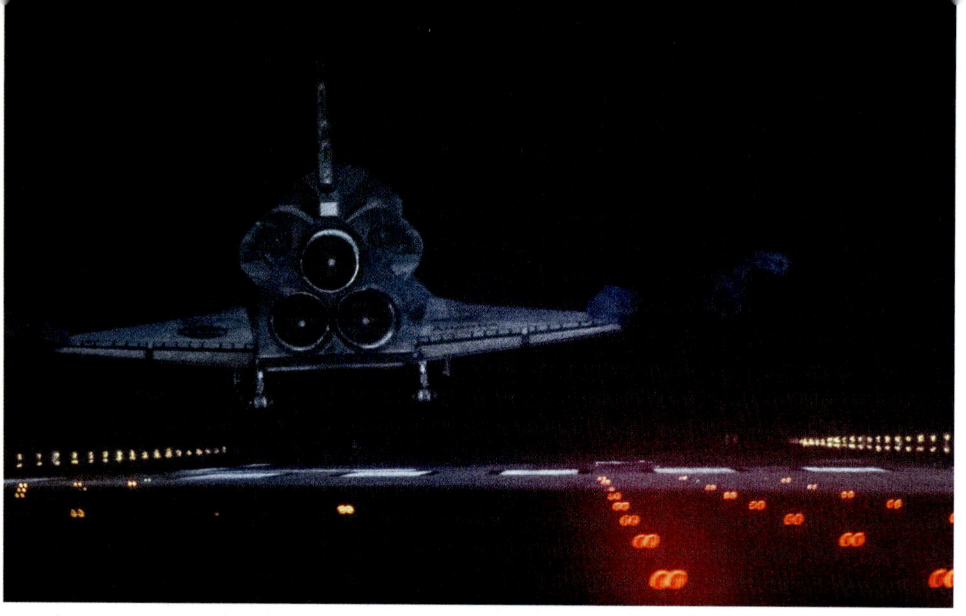

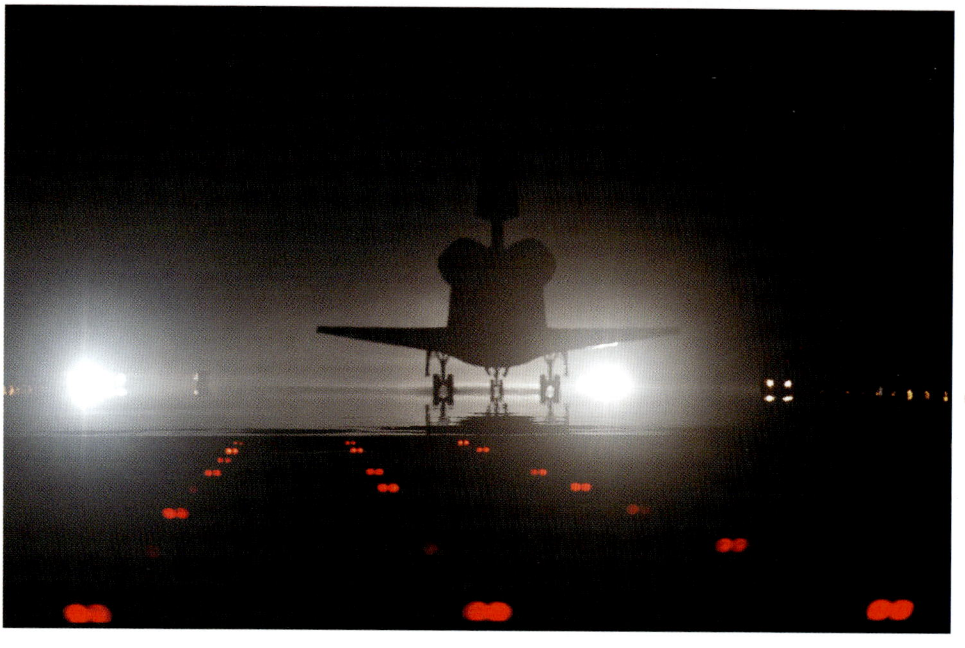

The Ultimate Challenge

(top and bottom) For me, the hardest part of a mission to photograph was a night landing. They weren't as common as daytime landings, but missions to the space station required very specific launch and landing times. When the shuttle comes back, there are no lights to track it in the sky, no noise from any engines, and the runway itself is lit with just two spotlights at one end. So when the orbiter approached the runway, you would hear a "swoosh" as air moved out of the way at more than 250 miles per hour, and then the shuttle would appear suddenly as a flash of dark in the lights. You had to be ready with the camera focused, and pan with it as best you could. Here, Atlantis completes the final mission in 2011, as seen from cameras looking from behind and down the runway toward the nose from the other end, a rare angle I wanted to capture before the program ended.

Endeavour Series

(this page) This series of shots of Endeavour on STS-134 capture the silhouette of the orbiter as it deploys its drag chute (to reduce speed), releases it, and then glides down the runway to a stop. From here, the shuttle would be put into a safe configuration and towed off the runway (as seen in the photograph on page 6) to begin preparing right away for its next mission.

Endeavour and Shuttle Carrier

(bottom) Sometimes, if the weather was too bad to conduct a Florida return, the shuttle landing would be diverted to California, landing at Edwards Air Force Base in the Mojave desert. This was the main landing site in the early part of the program, but became a backup later on. To get the shuttle back to Florida, it was mounted to the top of a specially-modified Boeing 747 known as the Shuttle Carrier Aircraft. During its cross-country return, which normally included at least one stop, the shuttle would fly at a low altitude and lower speed than normal commercial aircraft flights.

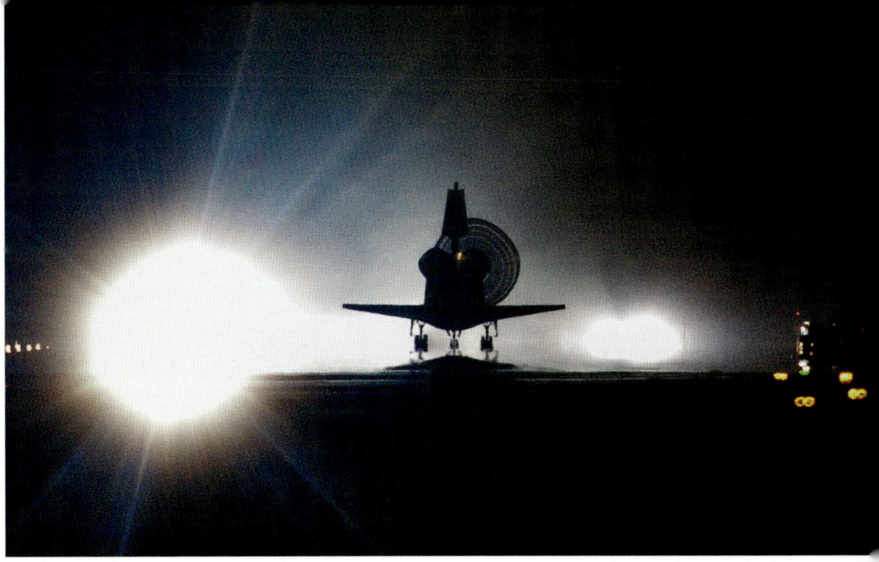

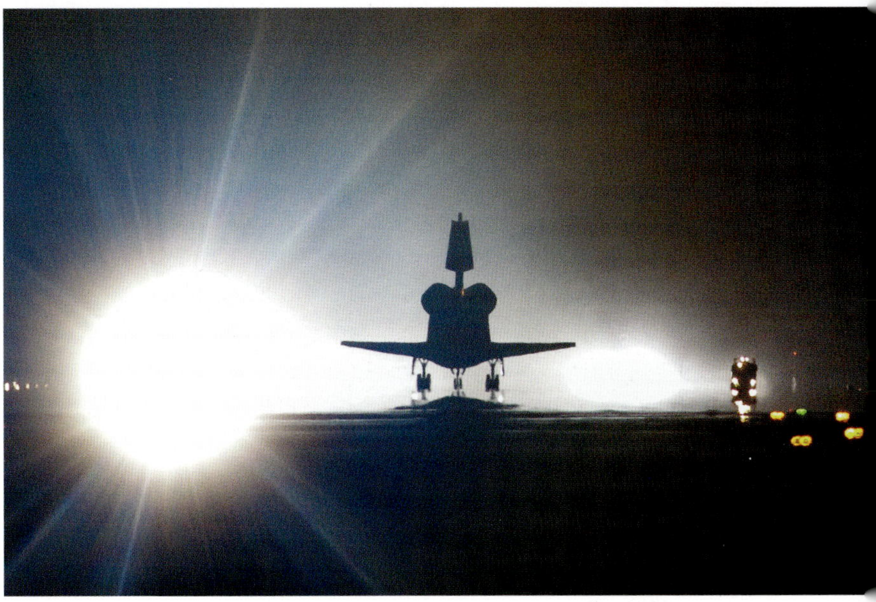

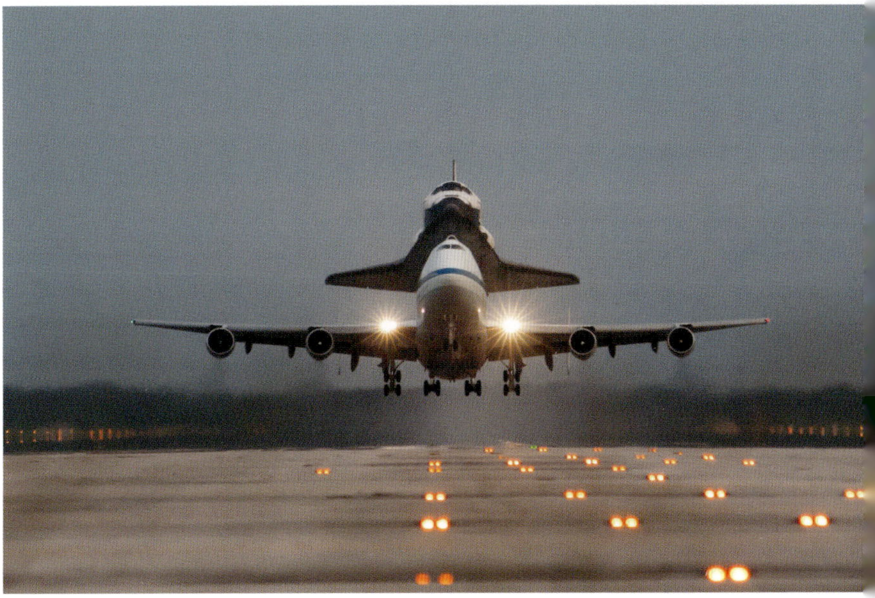

Crane System

Back at the Kennedy Space Center, this device, a crane system known as the Mate-Demate device, was used to lift the orbiter off of the 747 and back on the ground, where it would be towed to its hangar. In this image, Endeavour is being prepared not for its next mission, but for transport to California to be placed in a museum upon retirement.

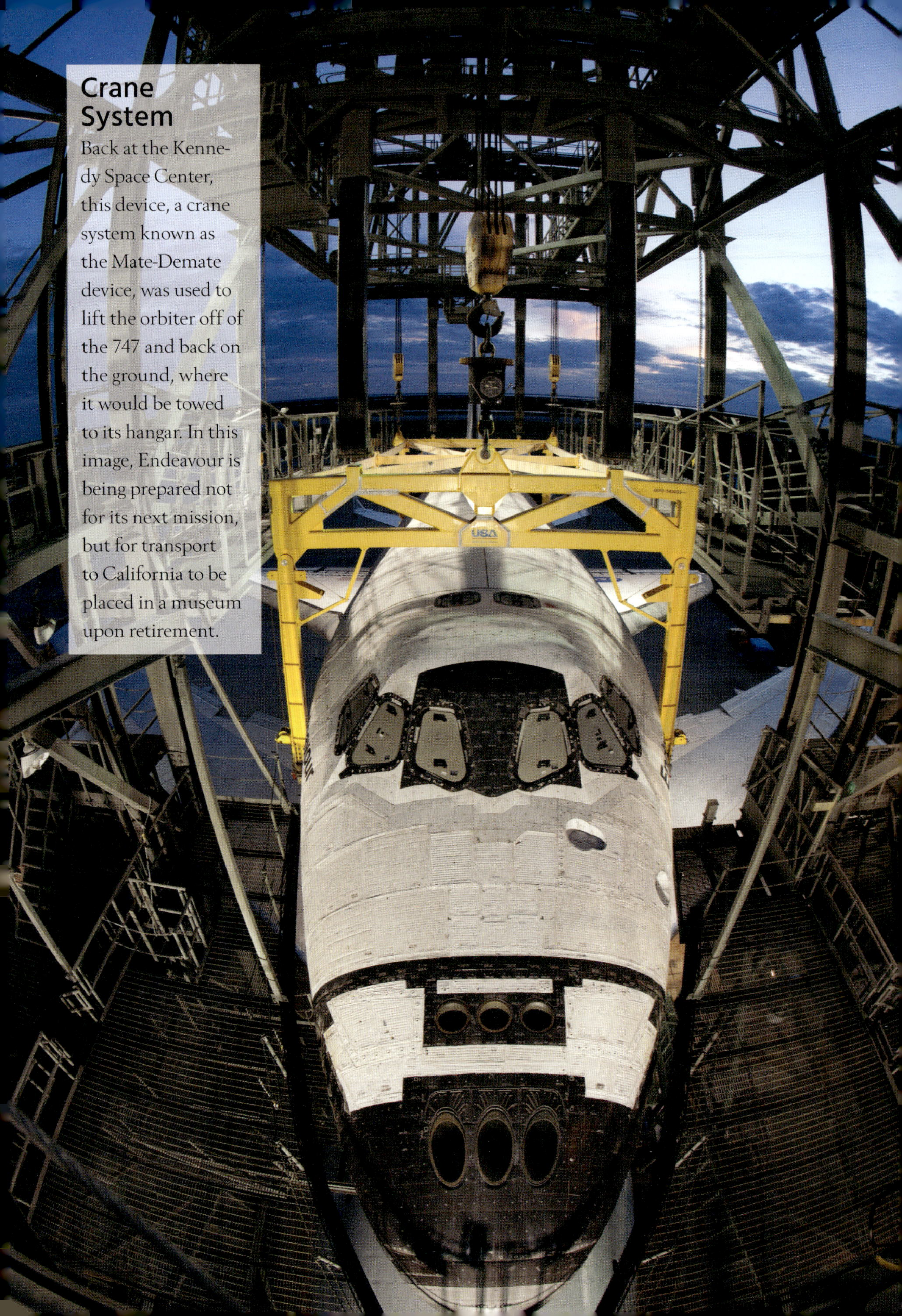

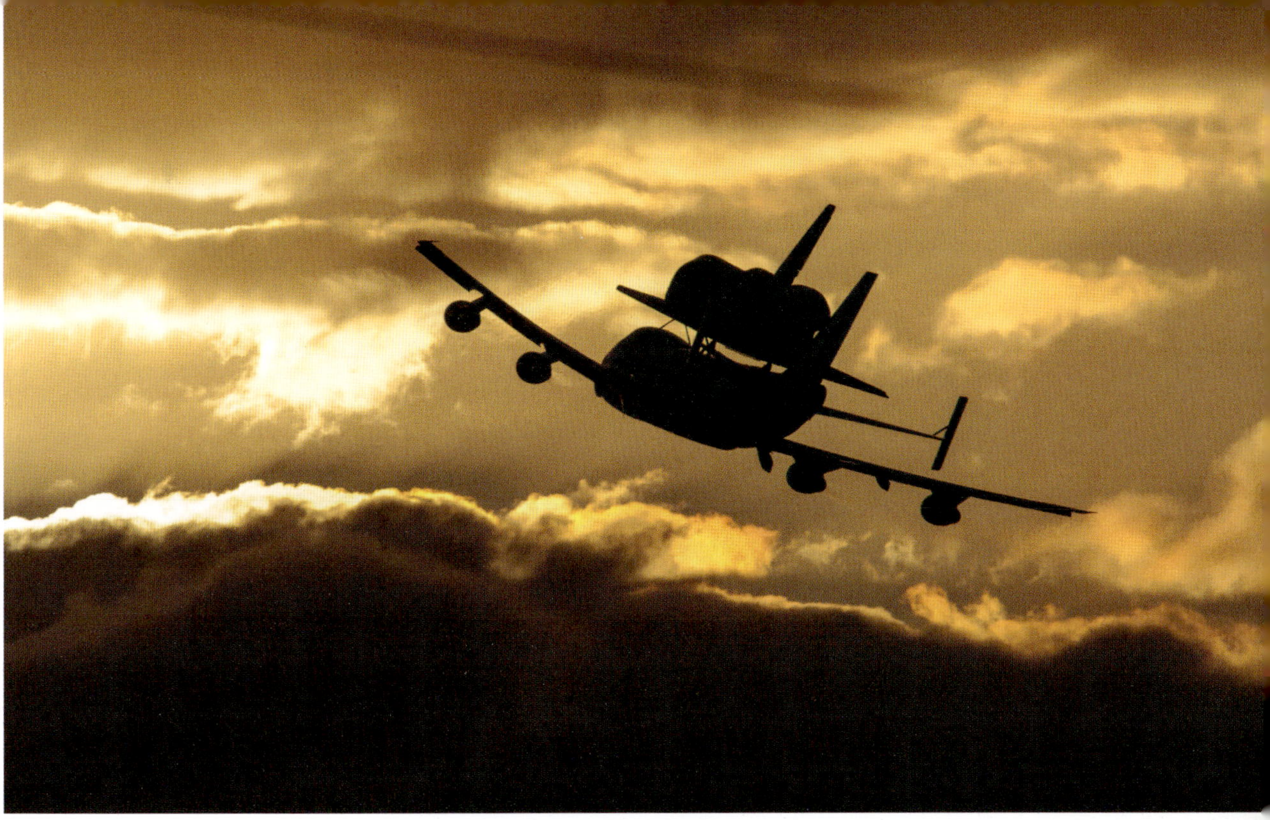

A Fond Farewell

(above) Discovery, mounted to the 747 Shuttle Carrier Aircraft, departs the Kennedy Space Center at sunrise in early 2012. This would be its final mission—a trip to the Smithsonian just outside Washington, D.C.

(right) The shuttles were all retired in 2011, Atlantis being the last to fly. While Endeavour and Discovery made their way to other parts of the country, Atlantis remained at the Kennedy Space Center. A fireworks display was conducted outside its future home at the local museum. It was the last time any shuttle would see the outdoors.

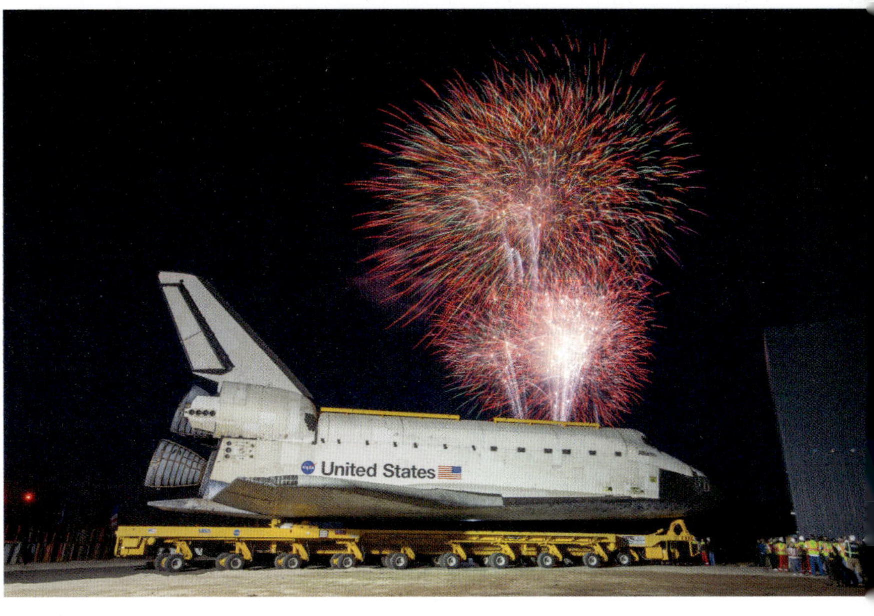

Flight Deck

(following page spread) I had the rare opportunity to photograph the flight deck of each space shuttle at the end of the program, and I wanted to get it just as it would be in flight, with the orbiter powered on and all the lights and screens lit up. This is one of the greatest and most complex machines humans have ever made, and it shows in this photograph.

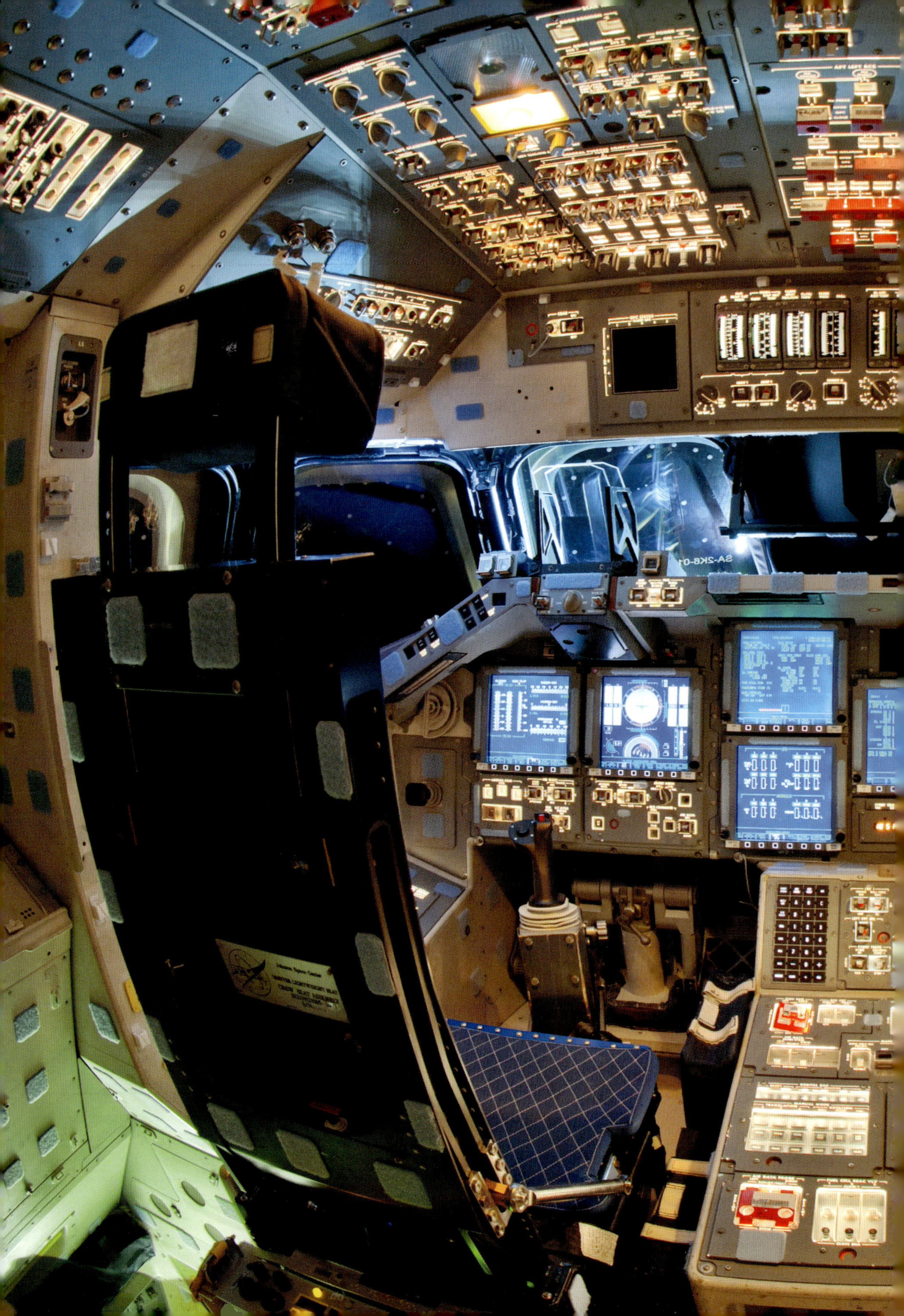

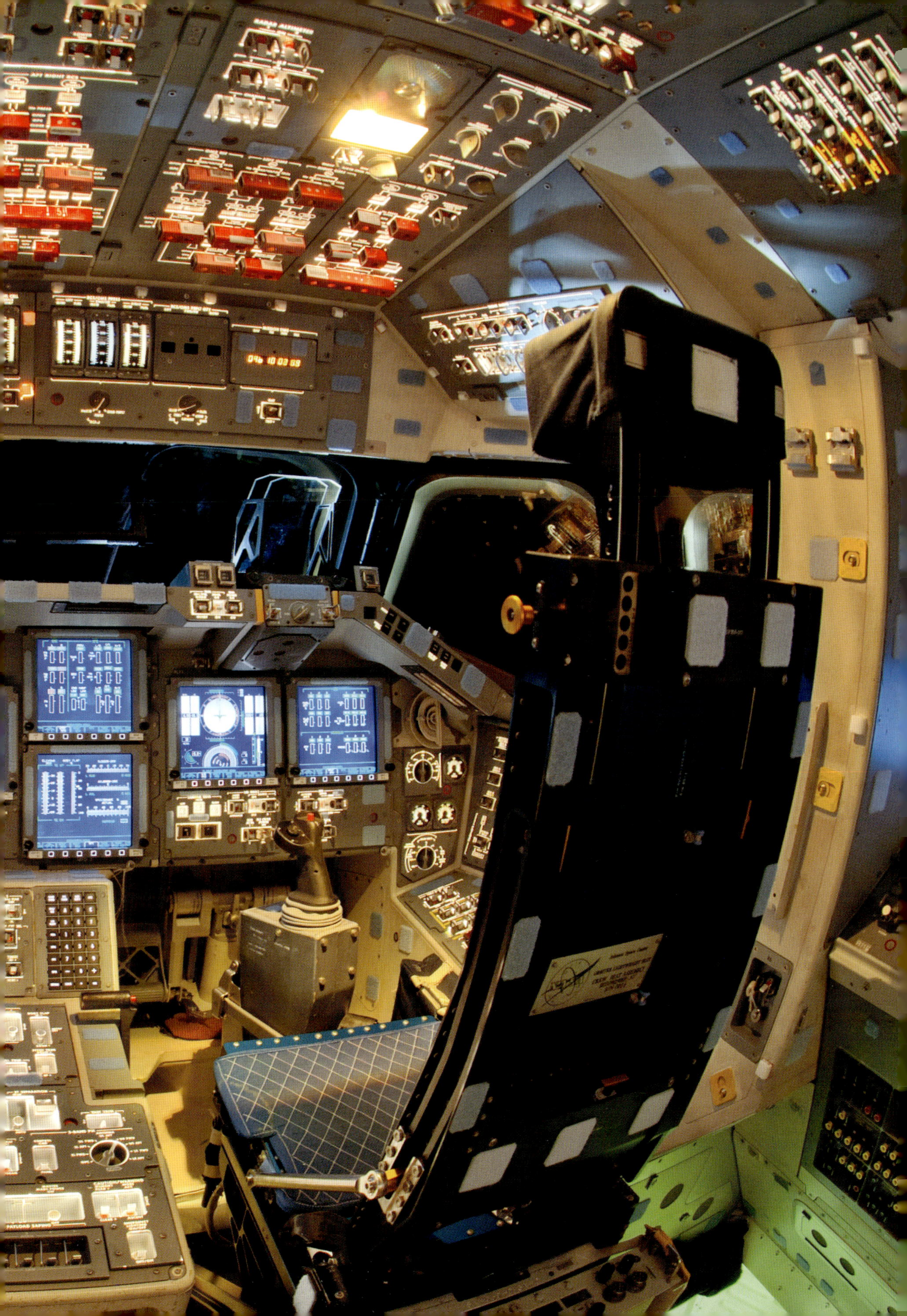

ROCKETS

Rockets are just as great an experience to photograph as the space shuttle was. For one thing, there are a lot more of them: different rockets launching off different pads, and at different launch sites around the world as well. In the U.S. alone, while the majority of space launches take place at Cape Canaveral, Florida, others take place from Vandenberg Air Force Base, California, and in recent years from Wallops Island, a NASA facility on the Virginia coastline. (On rare occasion, small rockets have also launched from Alaska.)

The rockets range from the long-time heritage vehicles such as Atlas and Delta, operated today by United Launch Alliance, to the modern converted ballistic missiles such as Minotaur, operated now by Northrop Grumman, to the newer SpaceX Falcon rockets, of which the first stage of the rocket can now return itself back to the launch site to be reused again.

Gone Viral

(following page) Several years ago, someone from a prominent camera and photo company contacted me about this very popular photo, which went viral on the Internet. I was asked how it was possible to capture the light range in this photograph. Did I combine multiple exposures on a moving image? When I replied, "This was taken on film," he said "Oh! did not think of that." I photographed on film from the age of seven, until the end of 2007, when this photo was taken. It was the next-to-last roll of film I've shot to date. (I spent several years photographing on both mediums at the same time, up until that year.) While it will never be exactly the same, the quality of digital photography is better than ever, and superior in many ways. That said, I do sometimes miss the challenge of getting it down perfectly on film, and working with various types and brands that yielded different results, such as color.

The Delta IV-Heavy rocket, seen here, is the world's largest operational rocket in terms of size, as of 2019, a title it has held since debuting in 2004. The all-hydrogen-fueled vehicle produces unique, almost transparent flames from its engines and produces a large fireball when they ignite on the pad, sometimes scorching the side of the rocket and making for some dramatic photos (see pages 88 and 107).

Countdown

(following page spread) Rocket launch countdowns begin with either a rollback or a rollout: Ten minutes of exposure time were needed to capture the rollback of the Mobile Servicing Tower (MST) at Complex 17 the night before launch of this Delta II rocket on a mission to the moon from Cape Canaveral. It was the last launch from what was the oldest active launch pad in Florida, having been in use since the 1950s.

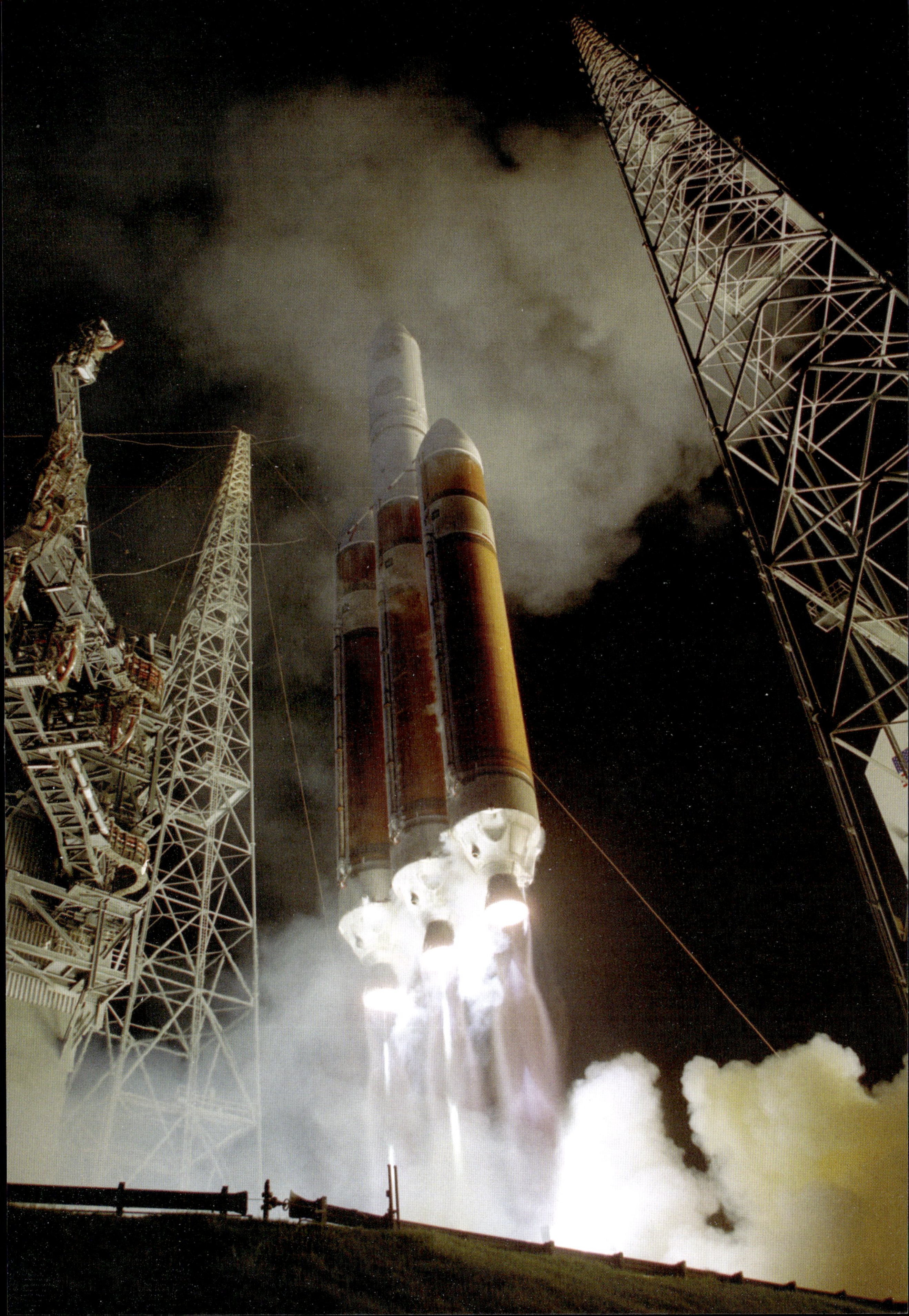

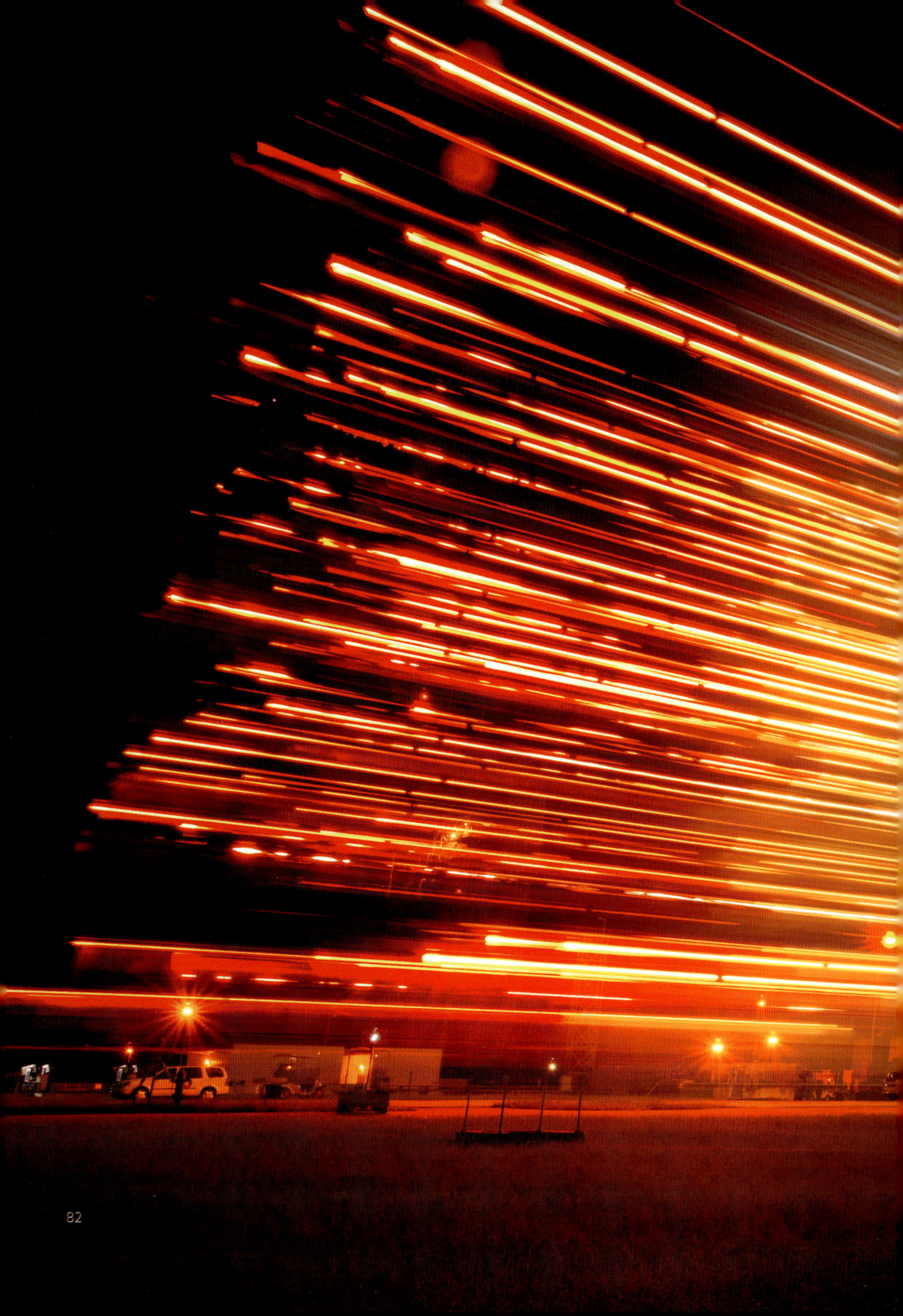

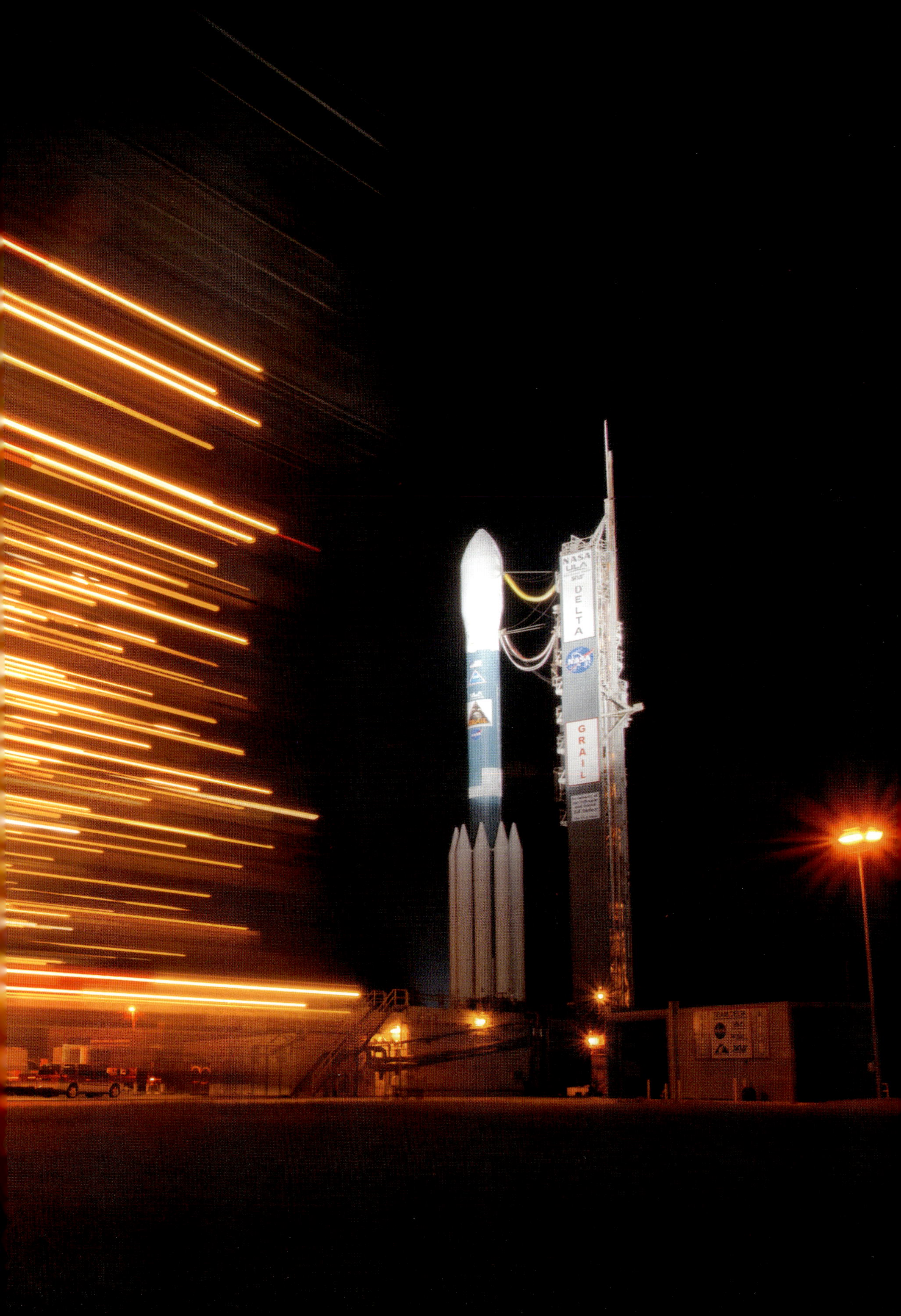
NASA
ULA
DELTA
GRAIL

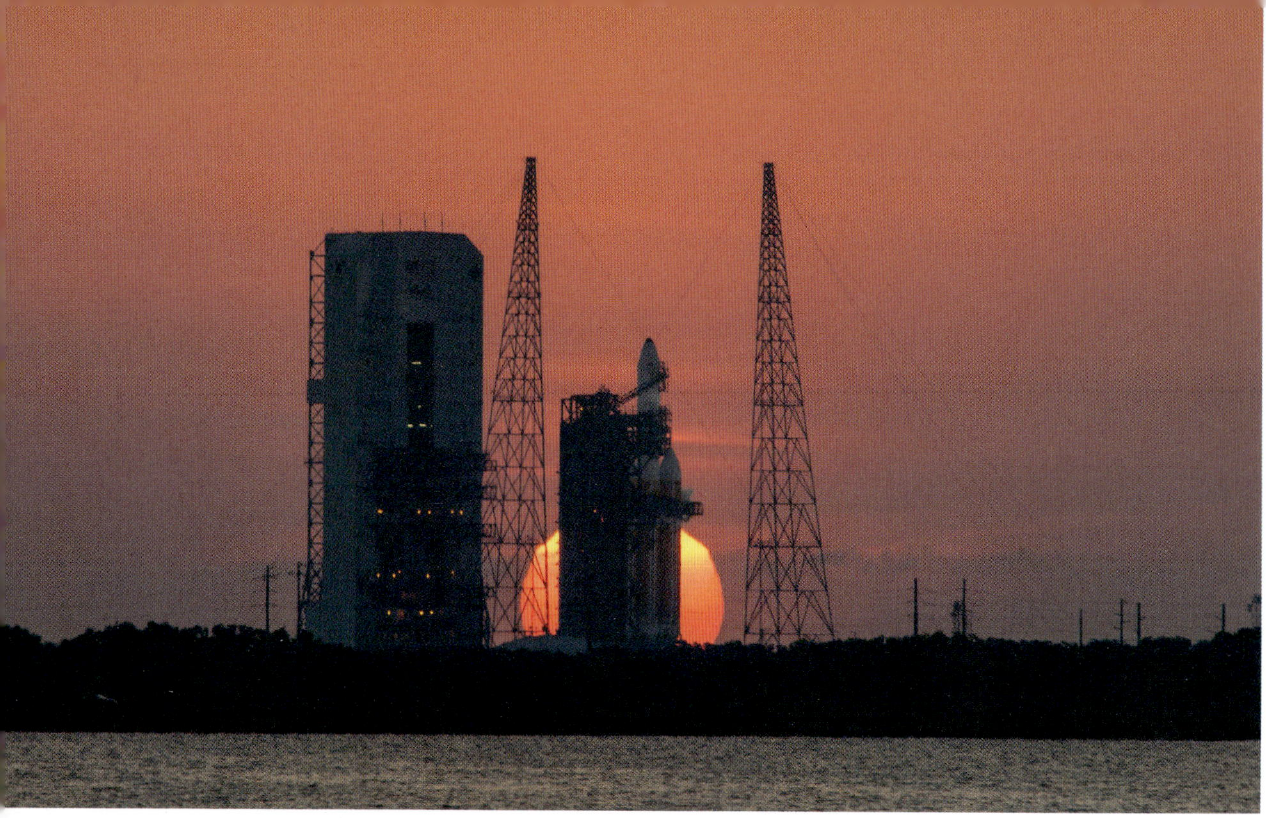

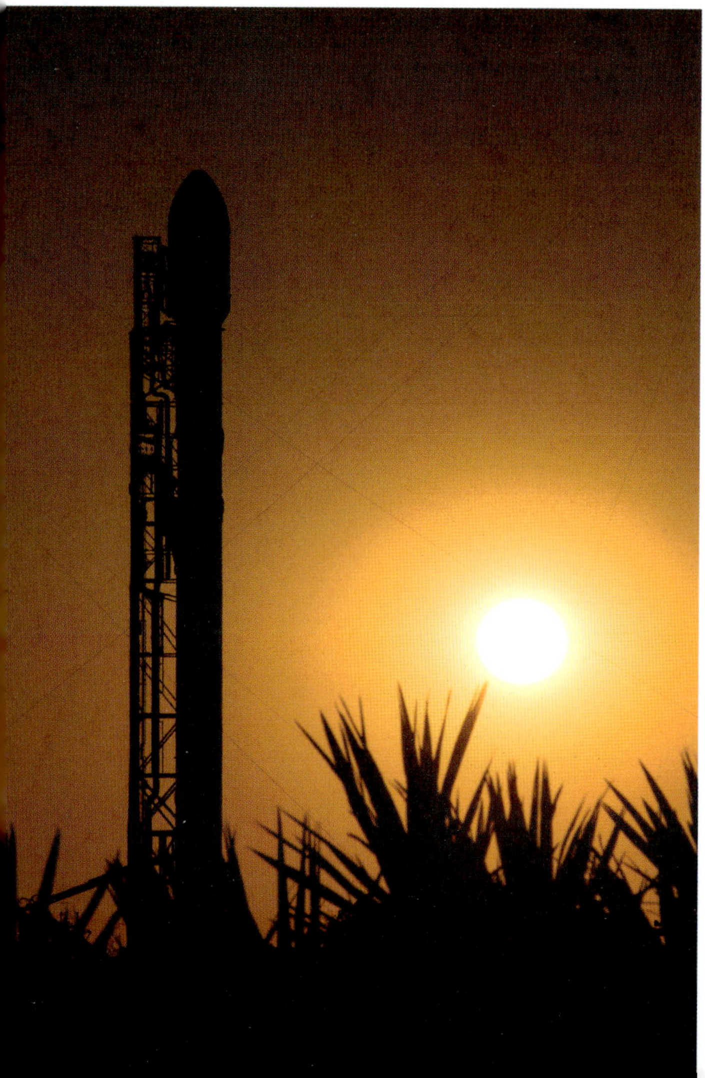

Sunrise

Capturing the sunrise *(above)* or sunset *(left and following page)* behind a rocket can make for a dramatic composition. Capturing the sun or moon behind a launch as it happens is a challenge, however. Launches can occur at any time of the day and are timed to precise minutes and seconds; the opportunity for one combined with the perfect aligning of both objects and good weather, as well as an accessible place to be at the time of liftoff, is extremely rare.

Above, a Delta IV-Heavy is nearing liftoff just after sunrise.

Sunset

(left and following page) A Falcon 9 at Cape Canaveral and a Delta II at Vandenberg AFB, California, await their middle-of-the-night launches as the sun disappears on the western horizon.

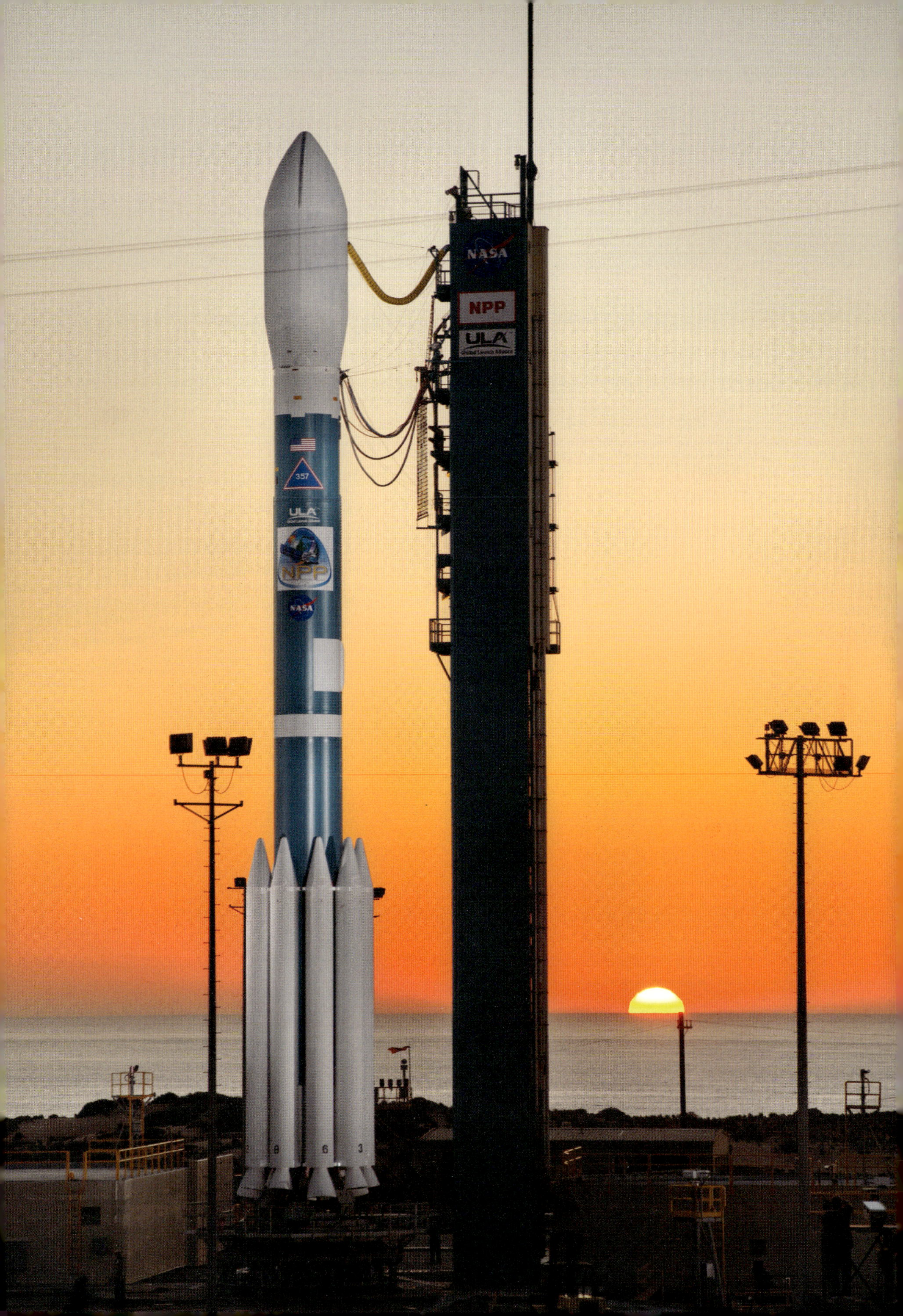
NASA
NPP
ULA
357
ULA
NPP
NASA

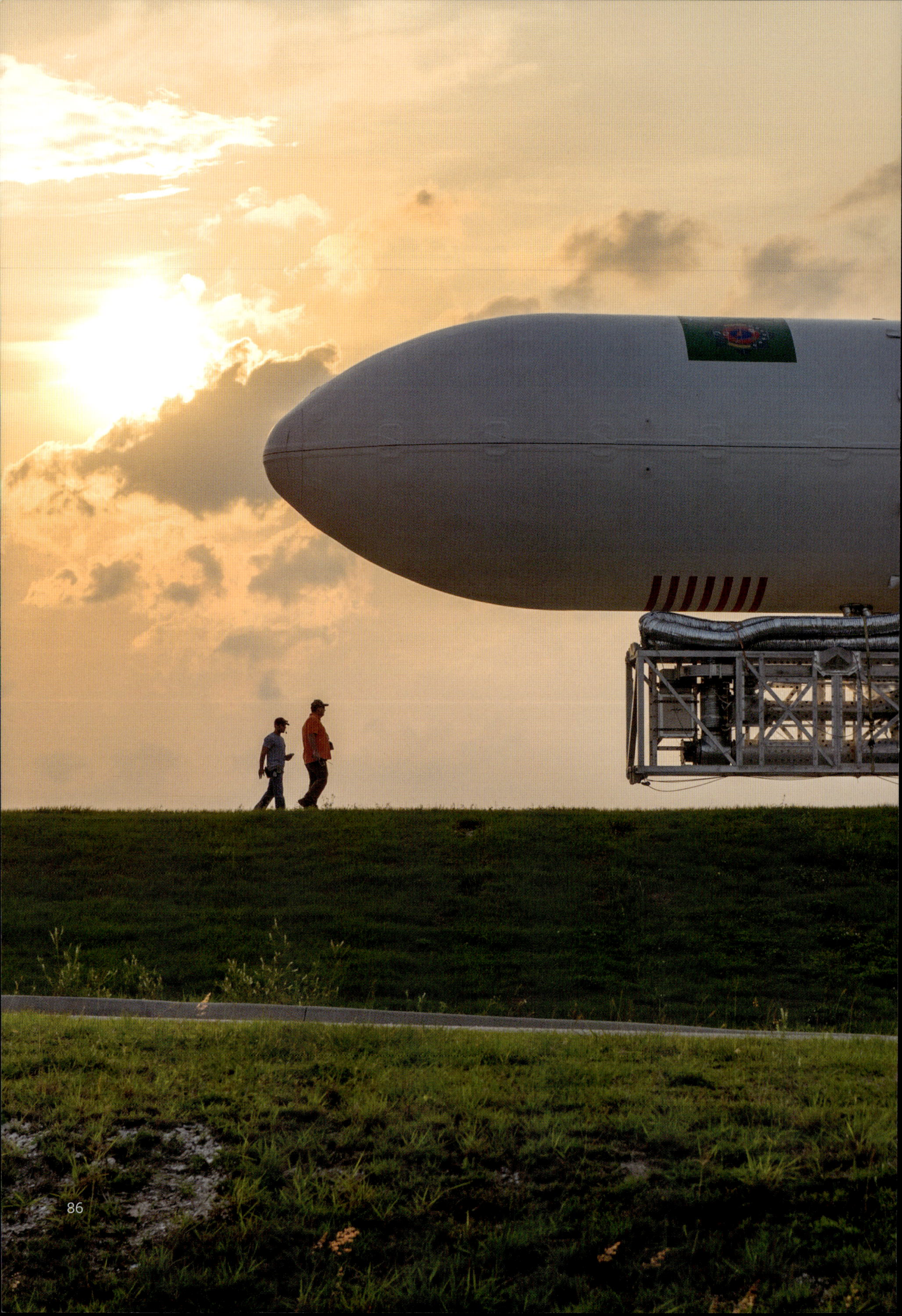

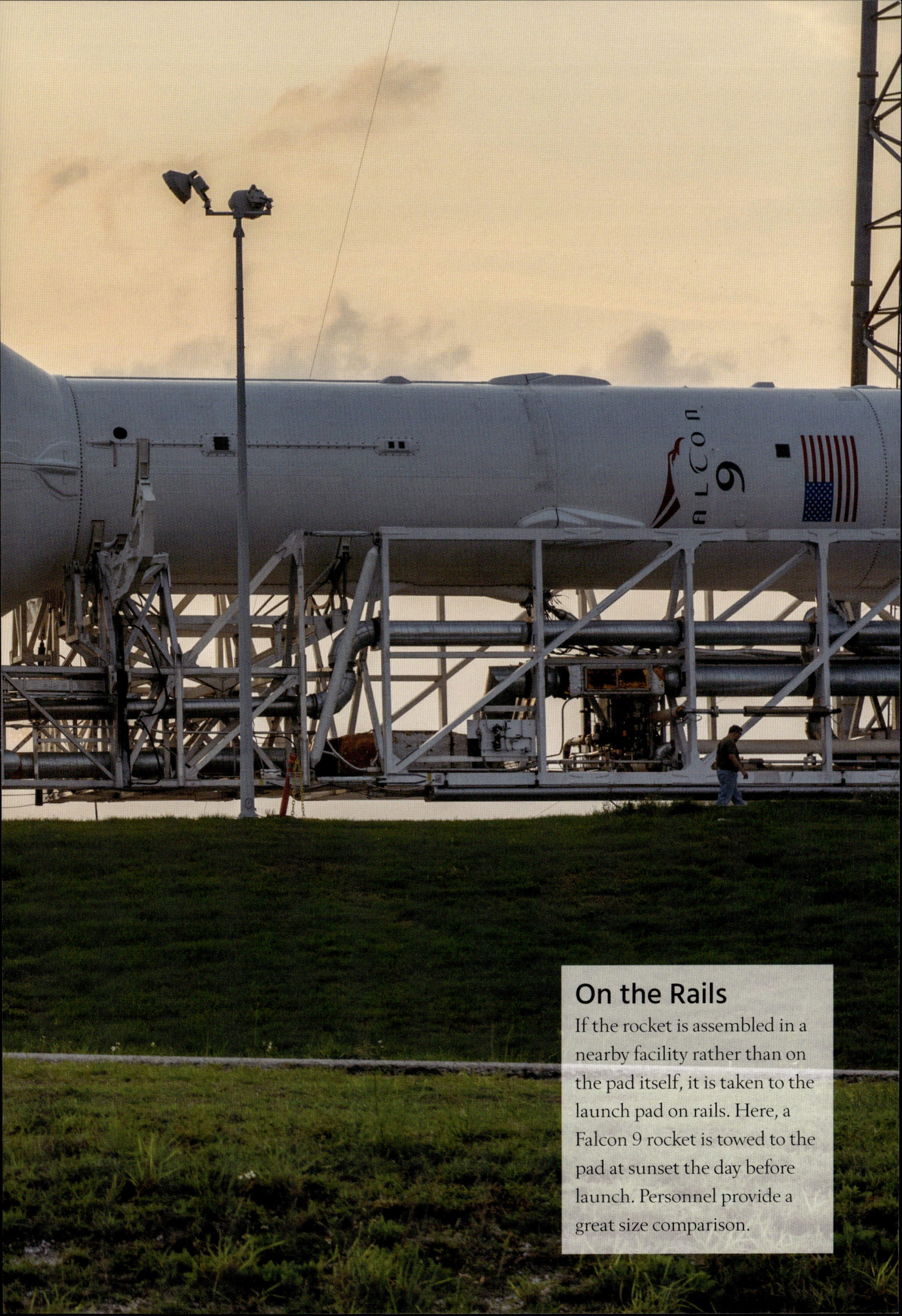

On the Rails

If the rocket is assembled in a nearby facility rather than on the pad itself, it is taken to the launch pad on rails. Here, a Falcon 9 rocket is towed to the pad at sunset the day before launch. Personnel provide a great size comparison.

The Moment of Ignition

(top) Because the majority of my cameras at the launch pad are designed to take photos when the sound from the engines begins, these are good examples of the first image to appear on the camera.

Keep in mind that the exhaust from rocket engines can travel at several thousand miles per hour; what you see here can erupt in a split second.

Right at the moment of ignition for Delta IV rockets, excess hydrogen gas around the engines is intentionally burned off, creating a fireball that sometimes scorches the outside of the rocket. There is nothing to worry about; this is as designed, and helps prevent an excess of the flammable gas from lingering.

Here, a Delta IV lifts off at twilight at Cape Canaveral, where deep blues complement the orange flames erupting around the vehicle.

Exhaust Plus Water

(bottom) On pad 41, exhaust from Atlas V rockets combines with high volumes of water (to dampen sound waves) that are poured on and across the exhaust duct, which then gets thrown from the pad at booster ignition.

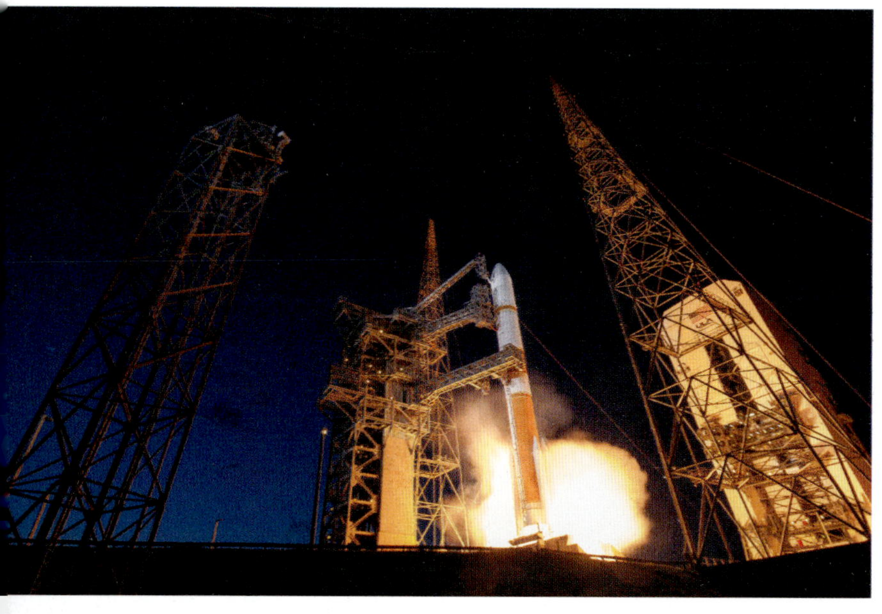

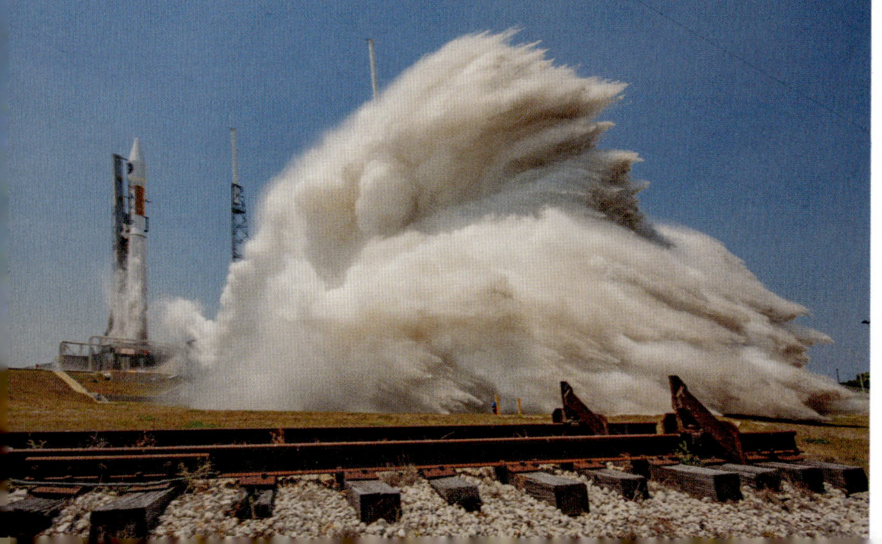

Engulfed

(following page) A Delta II rocket engulfs itself upon ignition of its solid rocket boosters on launch pad 17B at Cape Canaveral. This particular launch pad was unique; after undergoing a modification in the late 1990s, exhaust was sent harmlessly up as well as away.

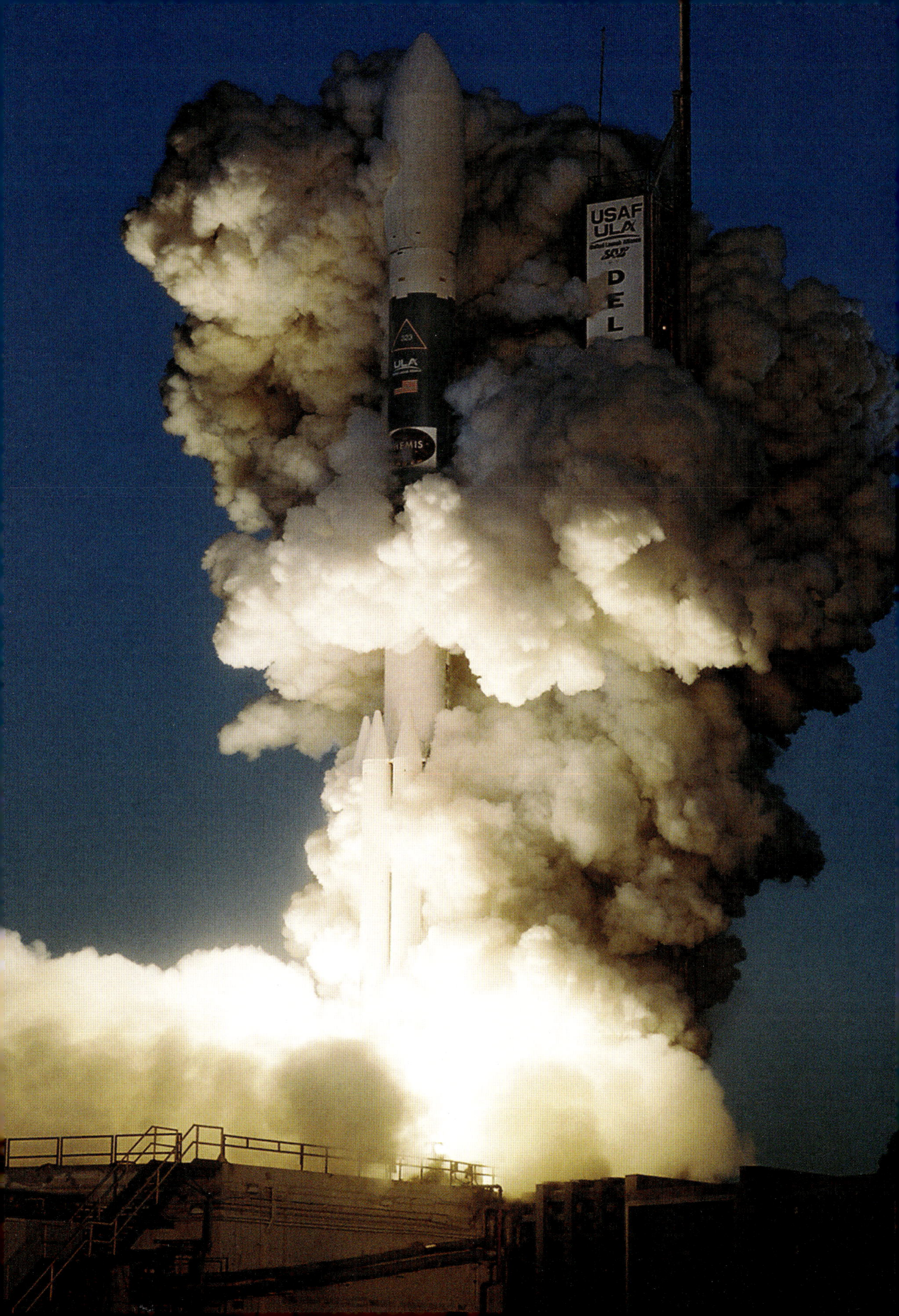

USAF
ULA
DEL
ULA

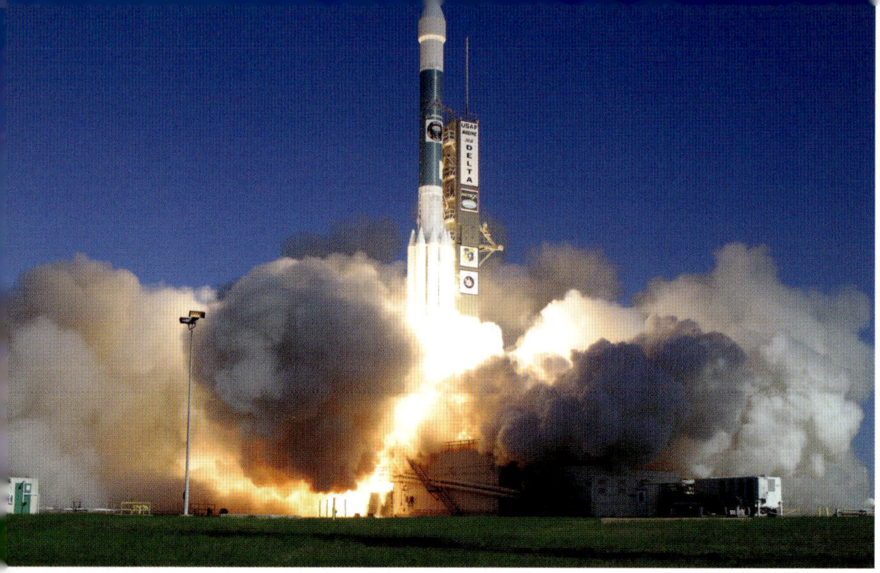

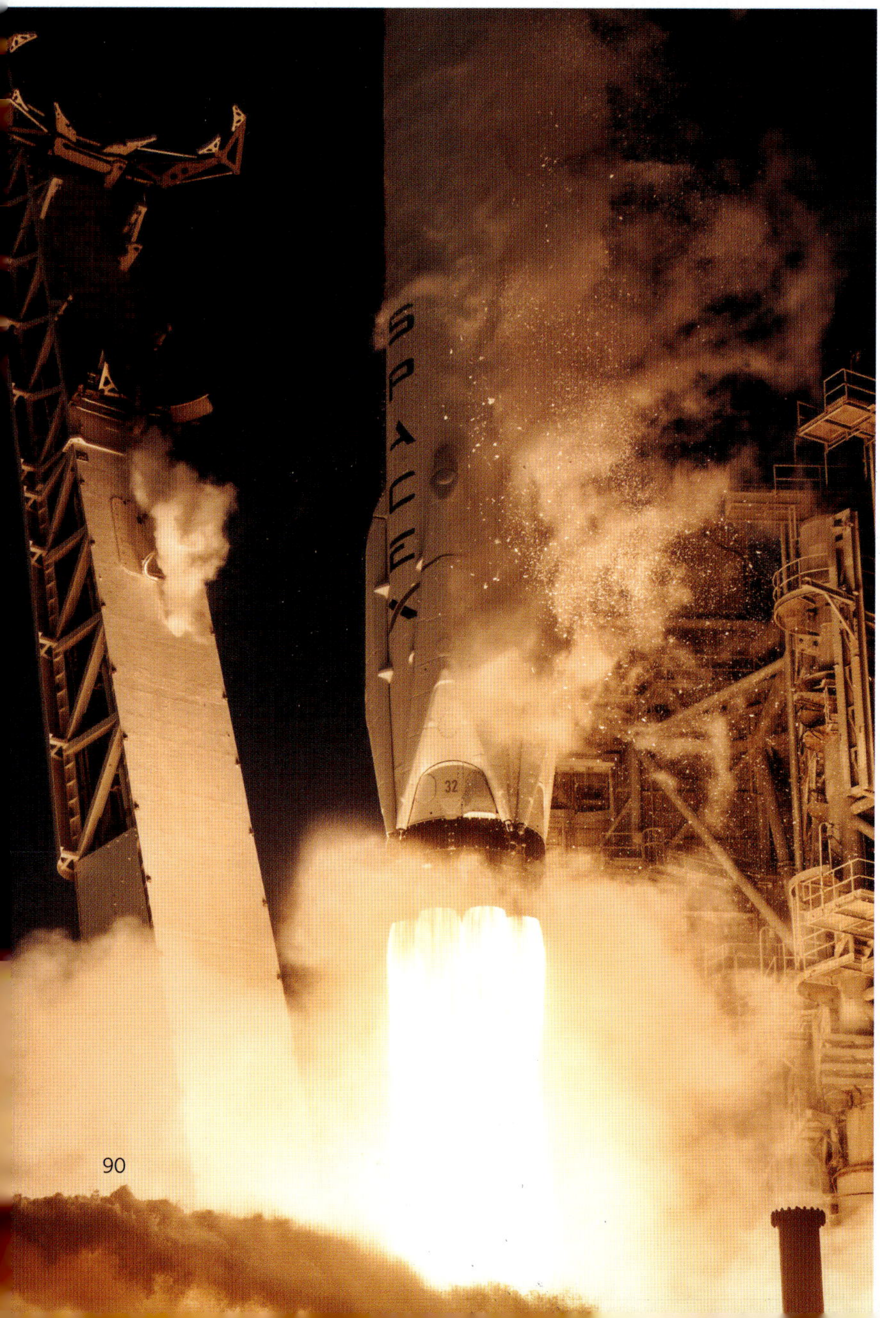

Delta II

(top) A Delta II rocket and six (out of nine) solid rocket boosters strapped all around the bottom of the rocket ignite. A cloud cast a shadow on the launch pad right before ignition. Sometimes, shade during daytime launches can make colors pop a bit more.

Closeup

(bottom) Closeup shots of the engines of the rockets have become more popular in recent years. Here, a Falcon 9 rocket, which has nine smaller engines on its first stage, has just lifted off near sunset.

Inside the Gantry

(following page) In this photo made from inside the assembly gantry, a Minotaur IV rocket has just ignited and lifted off. Using the gantry to frame a launch is something I like to do when possible.

Fire and Fury

(following page spread) Fire and fury explode from beneath this Minotaur IV rocket. Smaller than most other rockets in the U.S., and designed for smaller payloads, this vehicle was developed from an all-solid-fueled Peacekeeper missile first stage and takes off in a hurry, with no time sitting on the pad after engine ignition. When this booster ignites, it's going!

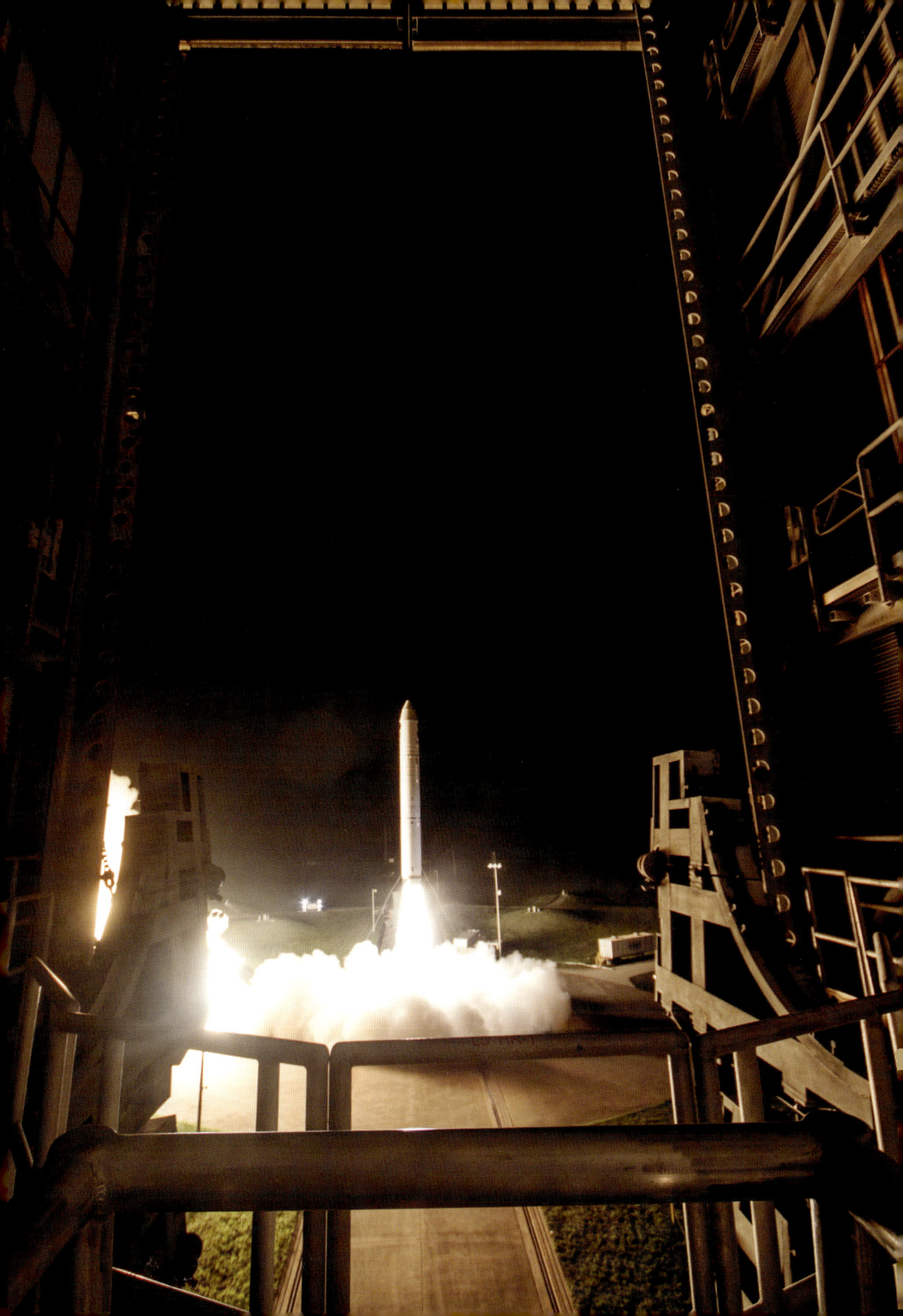

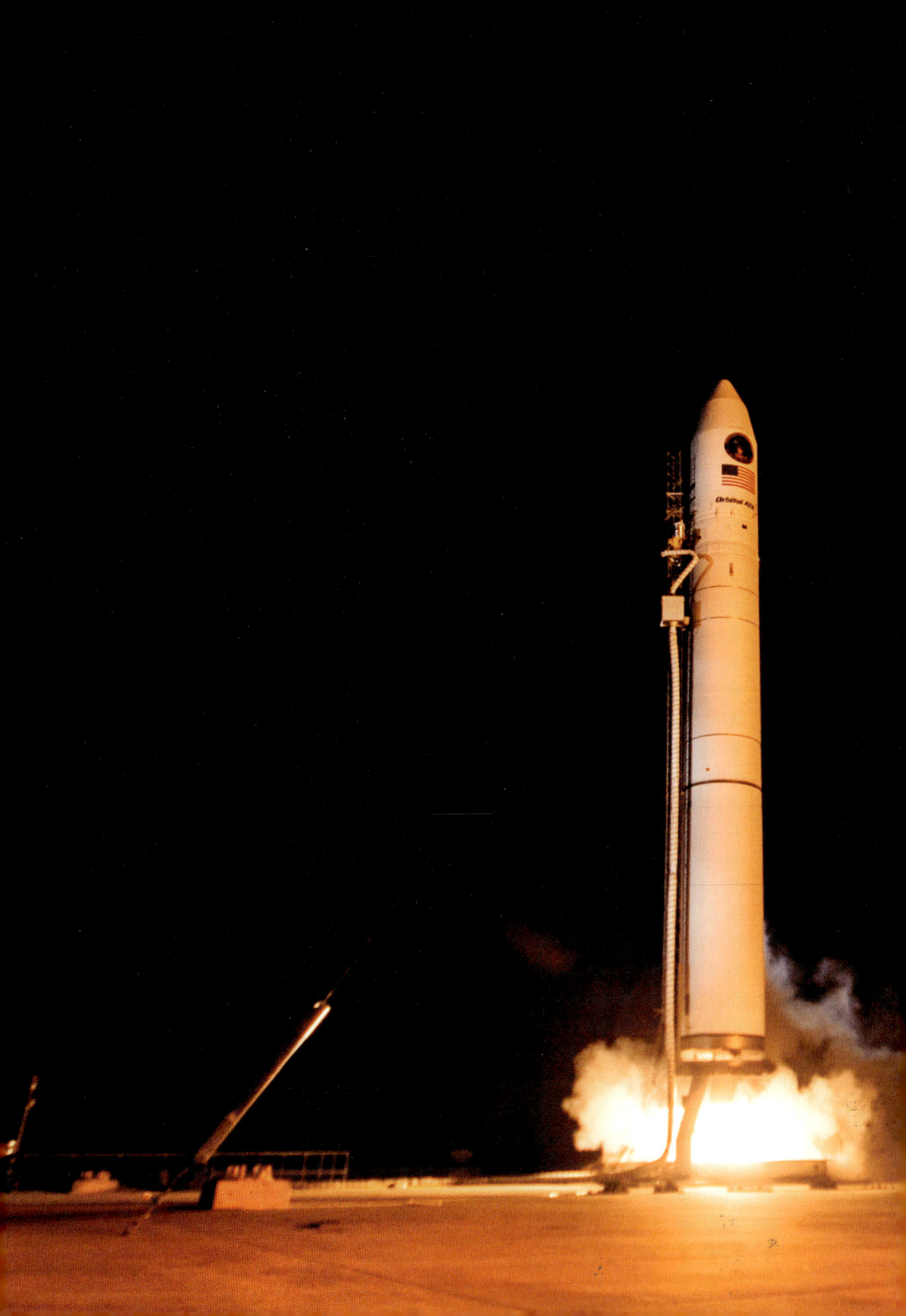

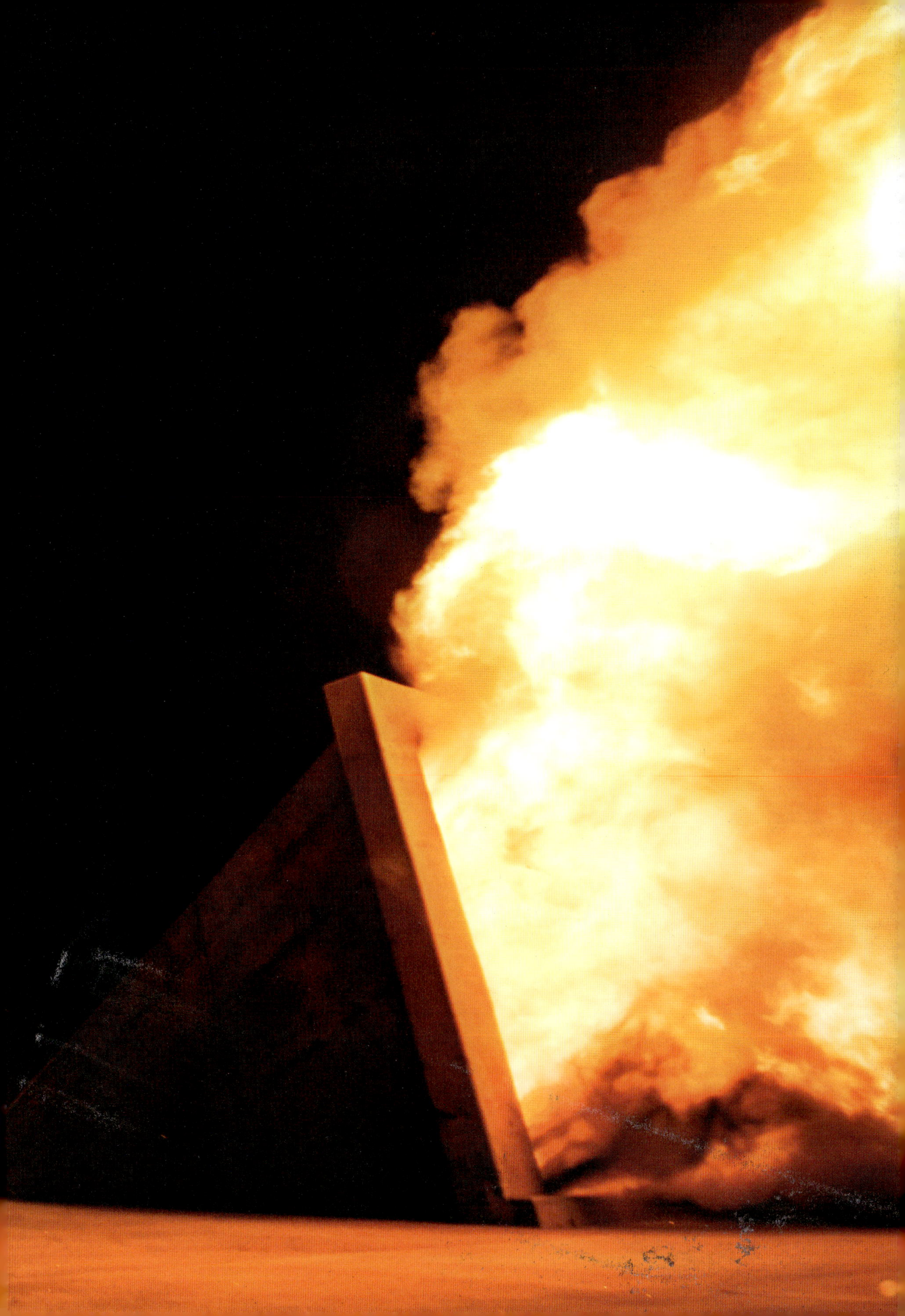

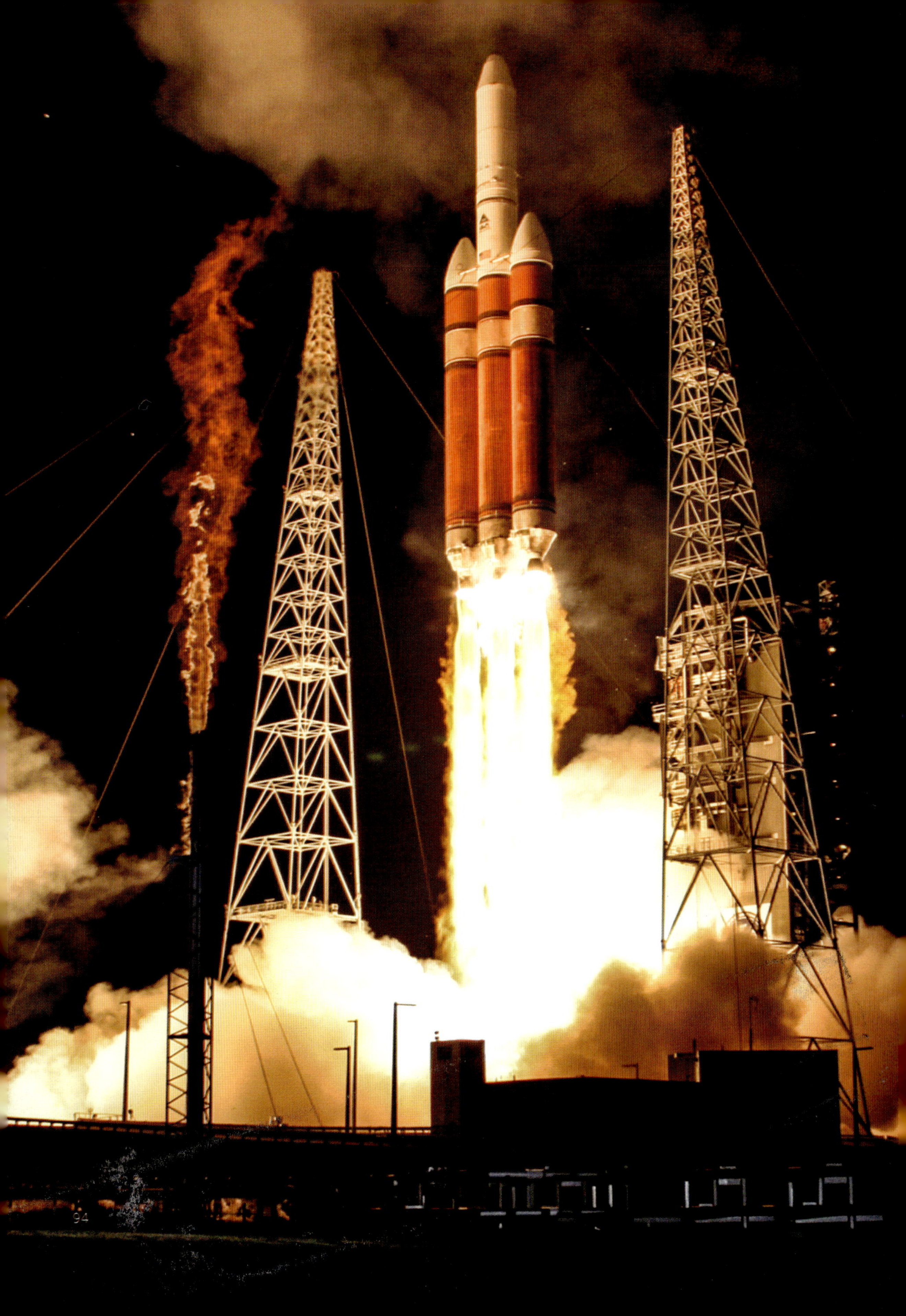

Safety Measures

(previous page) The United Launch Alliance Delta IV-Heavy is essentially three Delta IV rockets strapped together, side by side. Rockets such as this, that use hydrogen as their fuel, have a launch pad system that safely burns off excess hydrogen gas. Frequently invisible in the daytime, at night it can be a raging fire and provide a dramatic foreground subject for the launch.

Fish-Eye View

(top) A fish-eye lens placed close to this Atlas V launch captured the smoke exiting the flame duct at high speed.

Silhouette

(bottom) The setting sun silhouettes a Falcon 9 lifting off Pad 39a.

Careful Execution

(following page spread) For this shot of a Falcon 9 and its nine first-stage engines, the camera was located just 50 feet away and programmed to underexpose the flames by a lot. The image was captured in broad daylight, and what look like stars are ice crystals falling from the rocket.

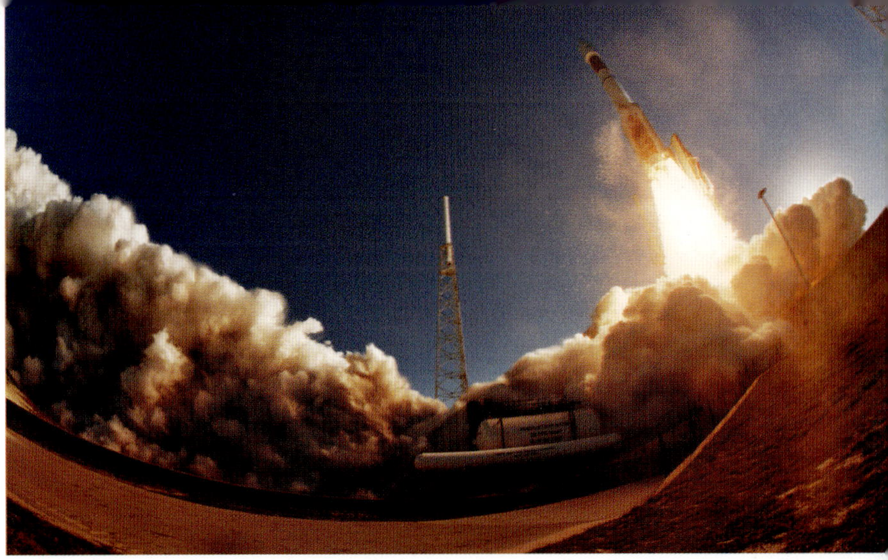

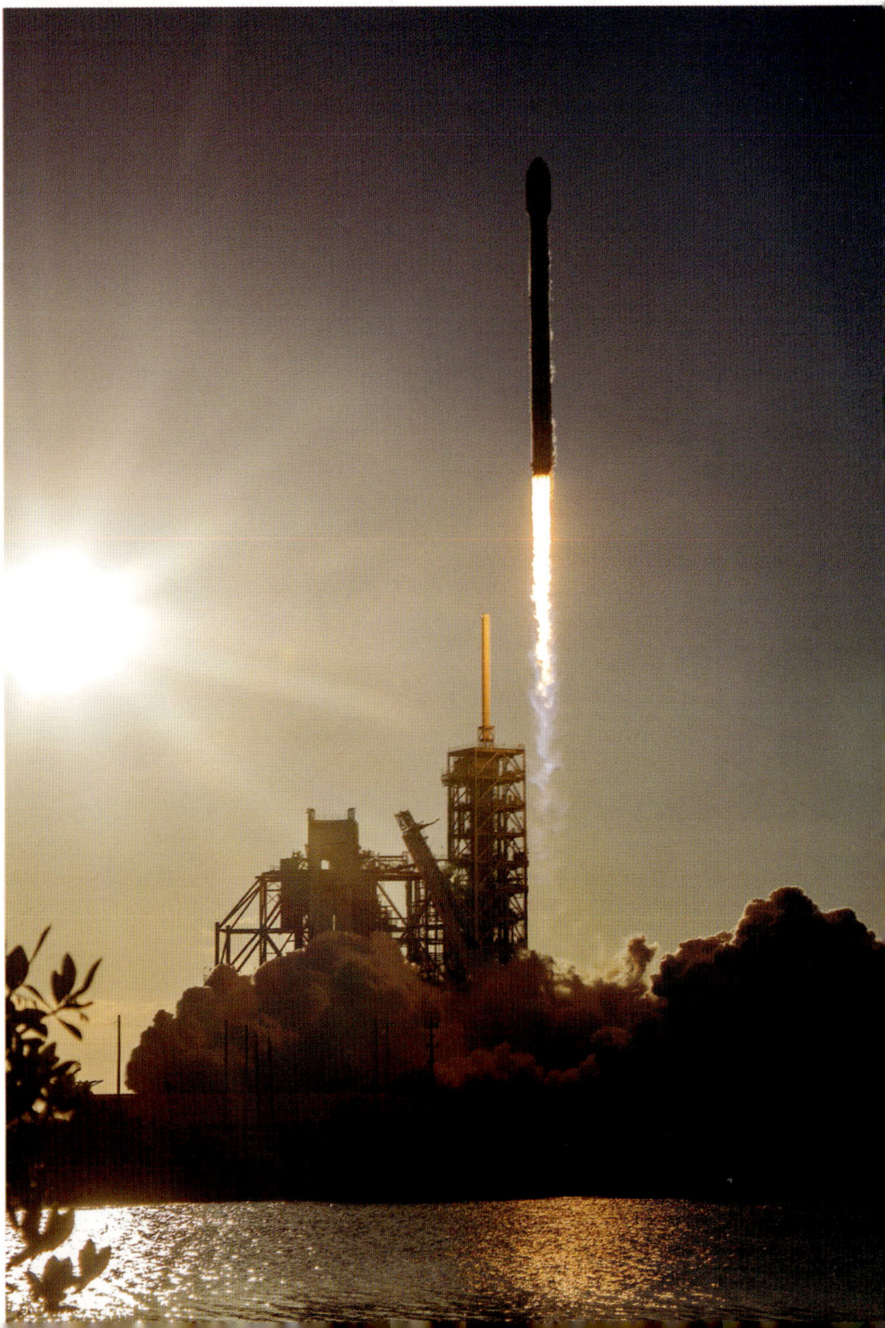

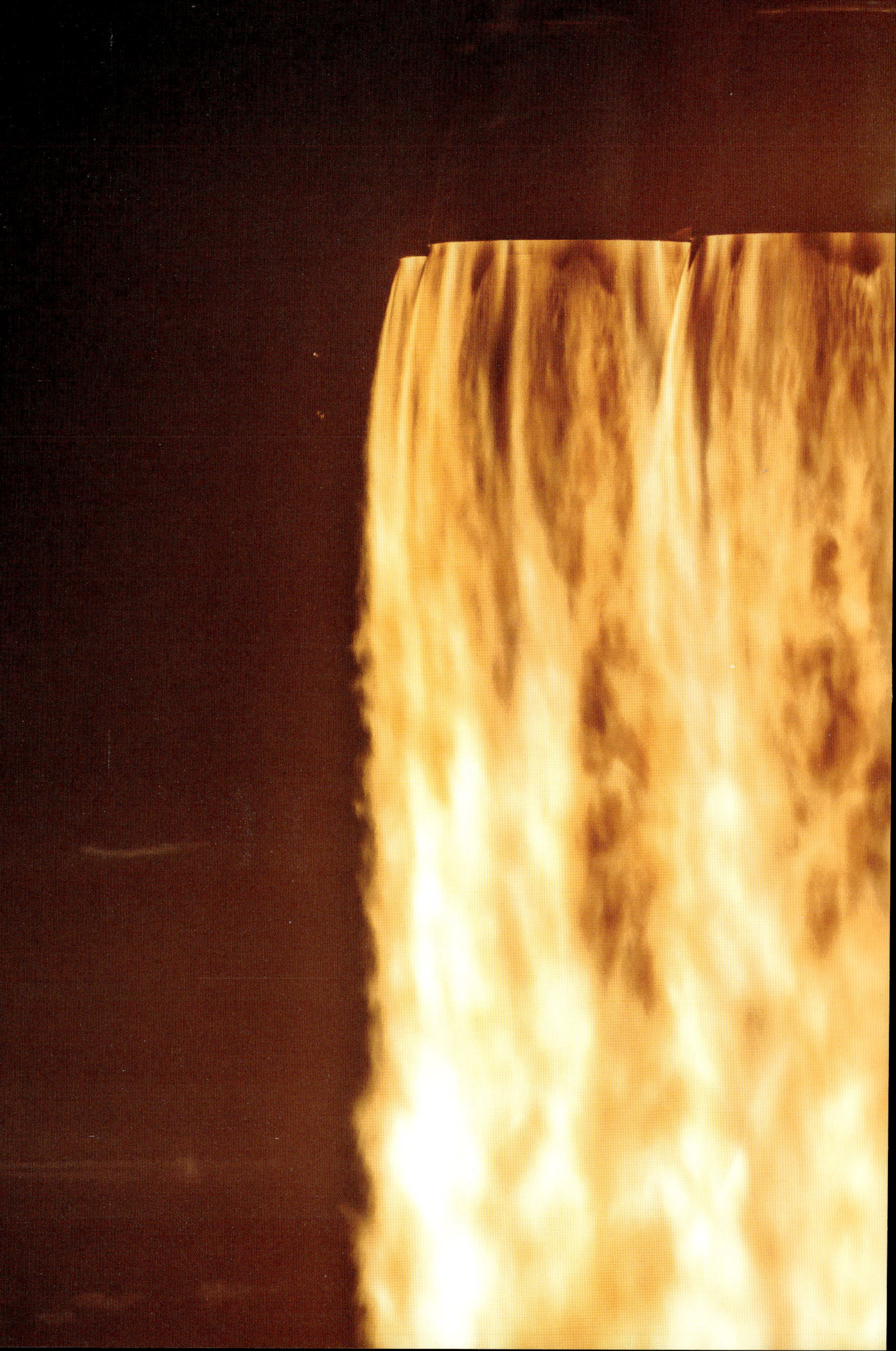

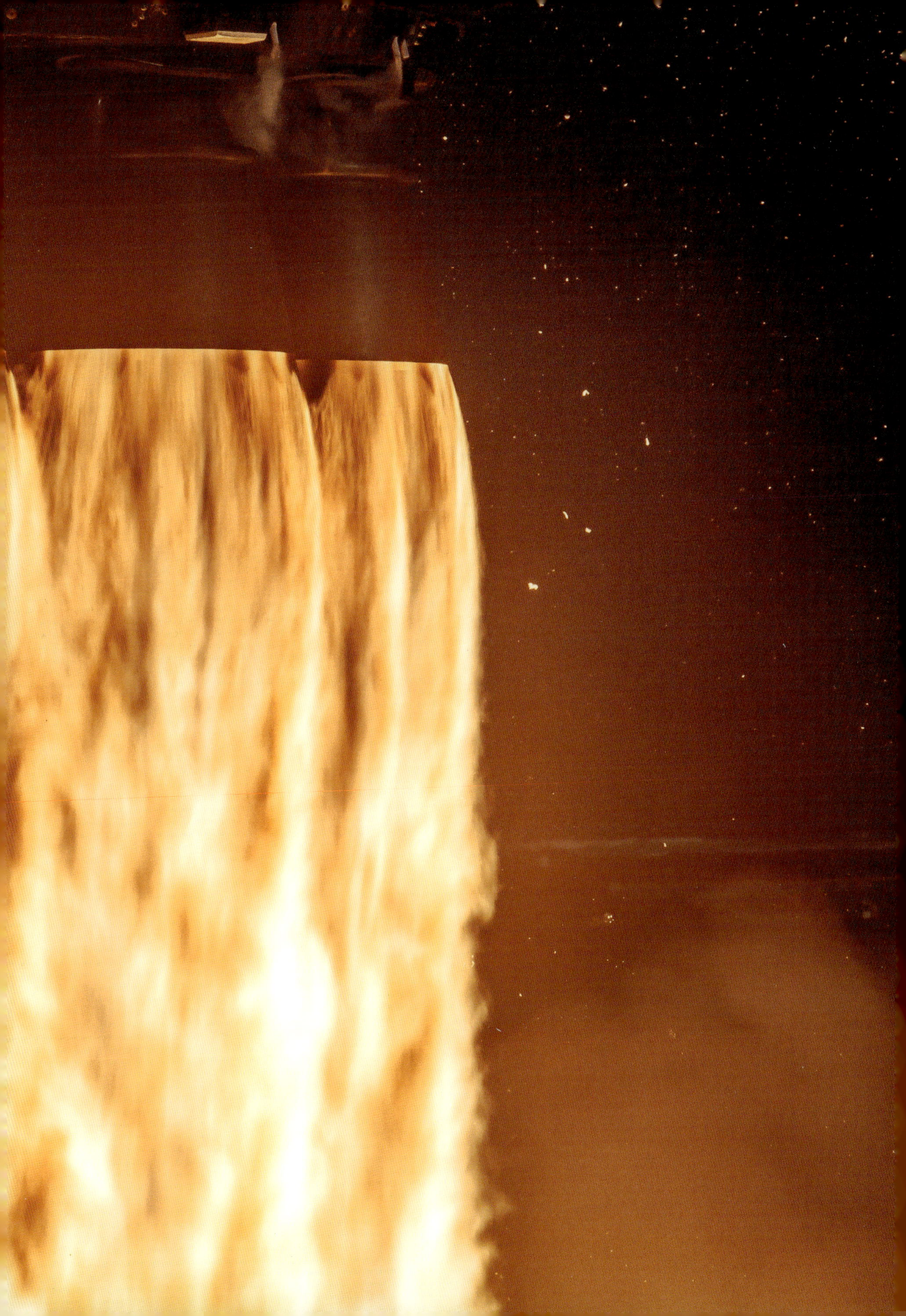

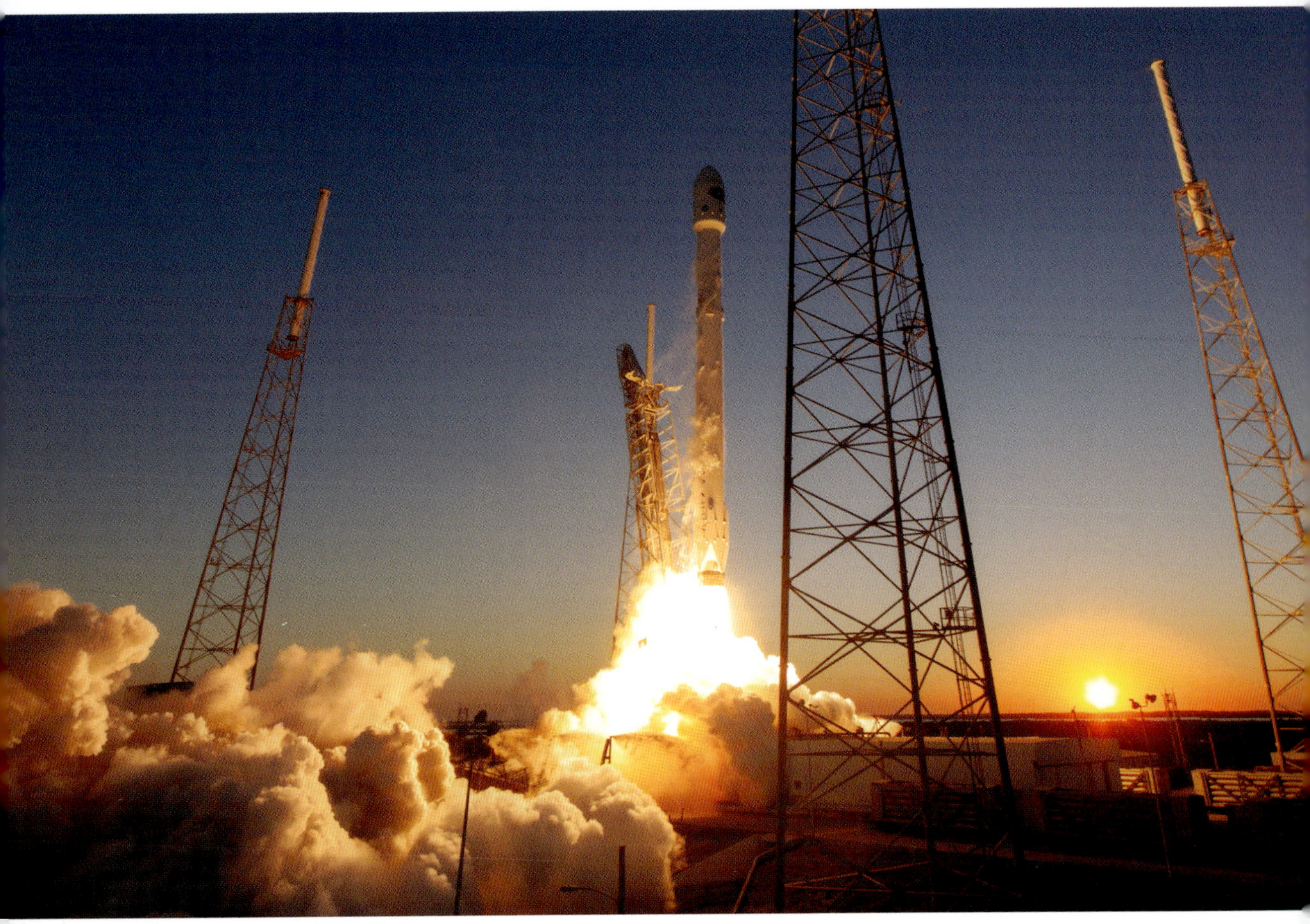

Falcon 9 at Sunset

(above) A gorgeous, clear winter sunset was the backdrop for this Falcon 9 launch of a NASA satellite. Launches such as this are rare (had this particular launch not delayed a few days due to bad weather, it would not have happened), and when the possibility arises, I do everything I can to get cameras, or myself, in a position to record the launch with the sun in the photo and composed properly.

New Horizons

(below) Silhouetted clouds provided a dramatic sky background for NASA's New Horizons mission aboard an Atlas V when it was launched on the world's first mission to Pluto in early 2006. In reality, the rocket is heading straight up, not at an angle. A wide-angle lens distorts the view, but adds to the drama. My favorite style of launch photography has, without a doubt, always been wide, dramatic shots like this.

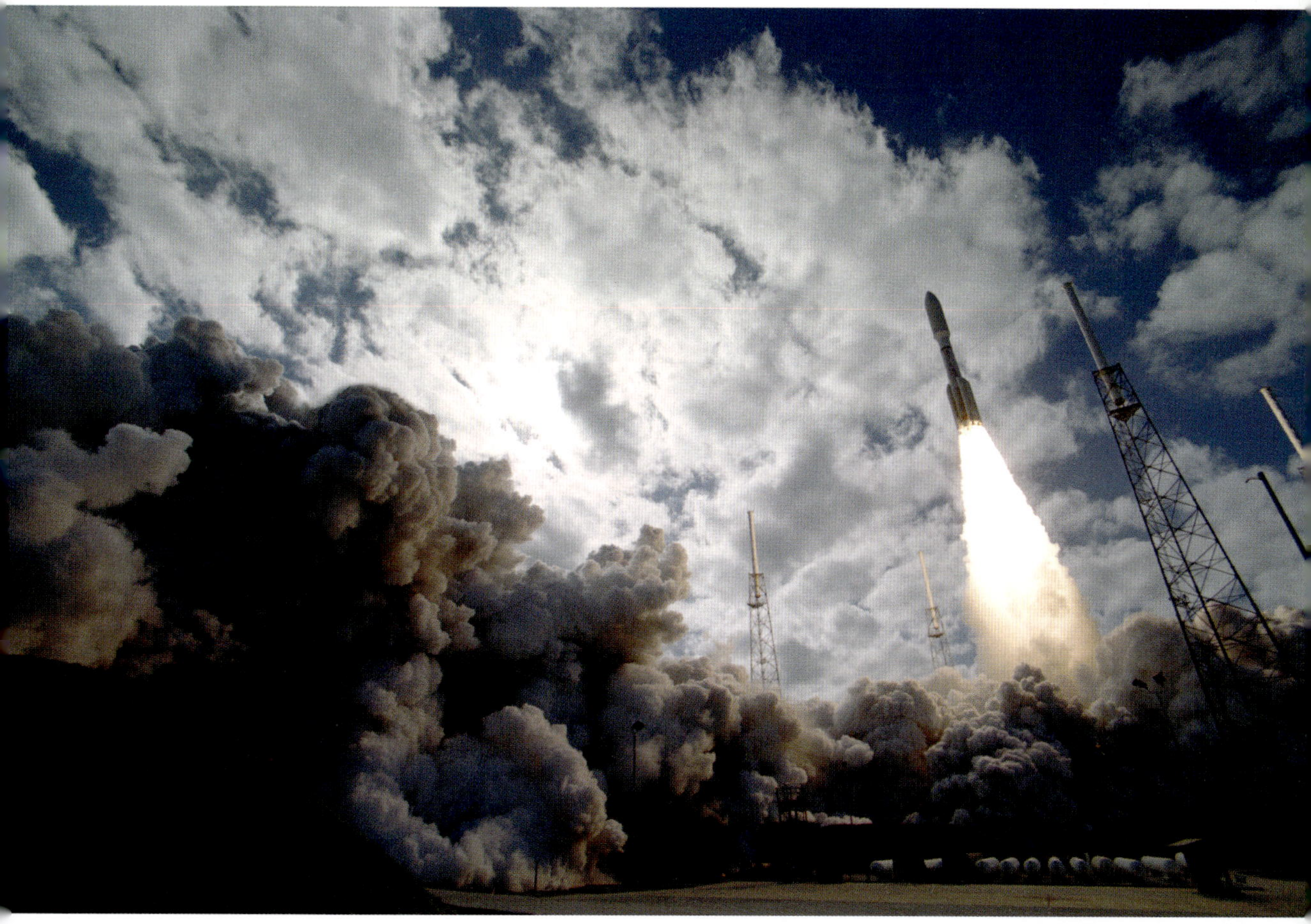

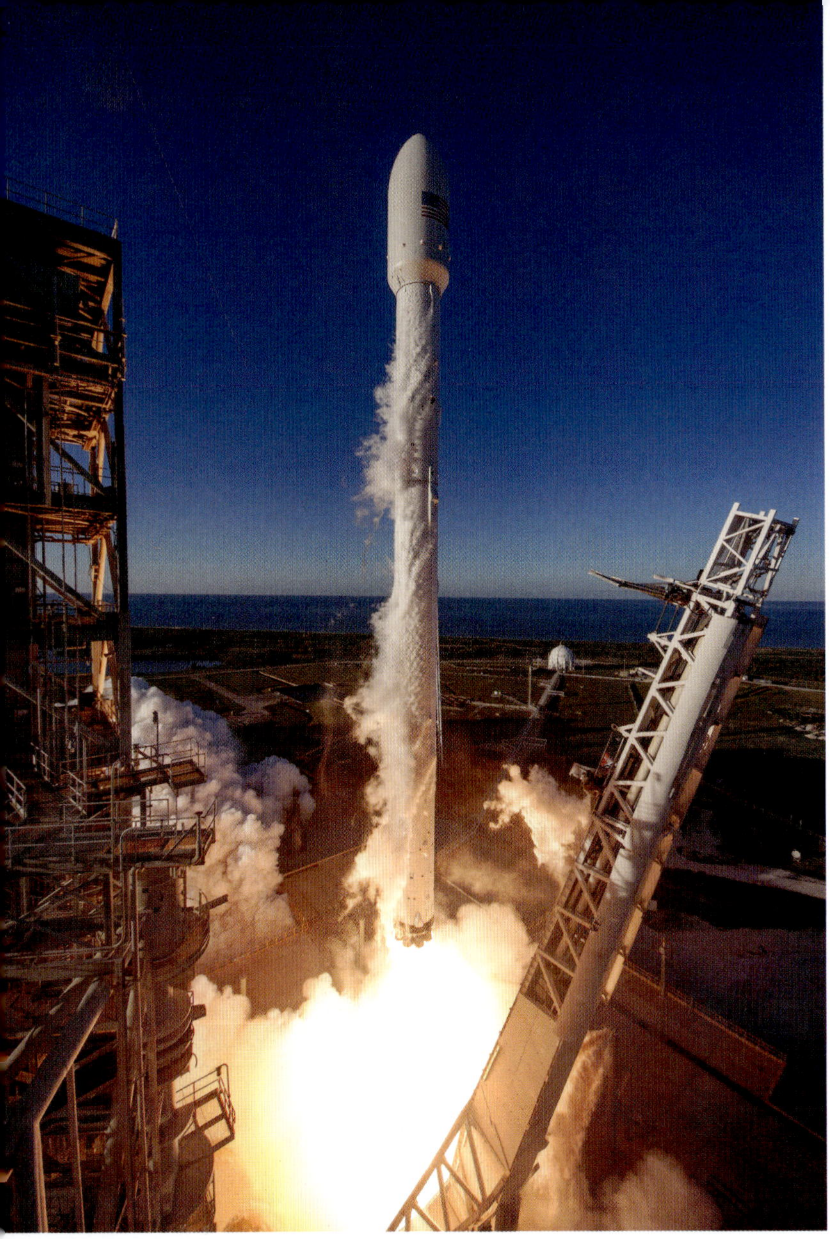

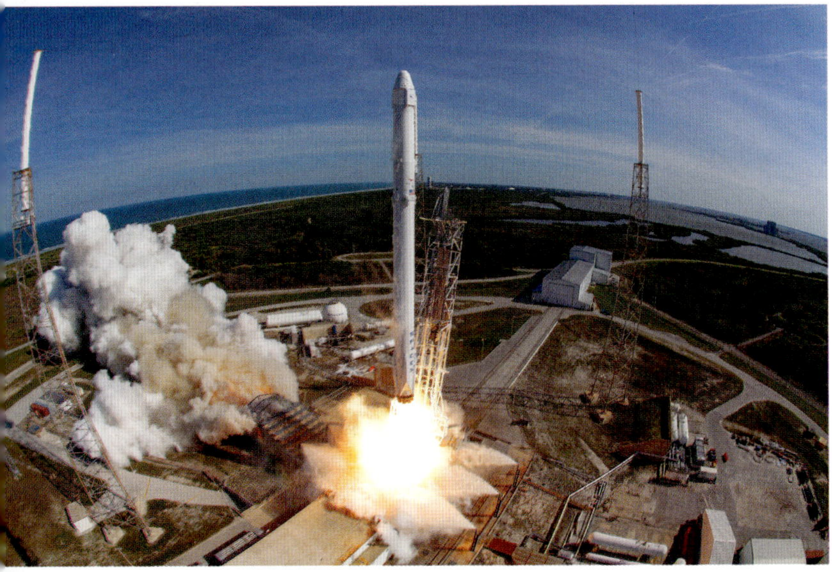

A Vintage Look

(top left) Falcon 9 begins a trip into space off pad 39A at the Kennedy Space Center. This angle is no longer possible, as the shuttle-era structure the camera is mounted to has since been removed.

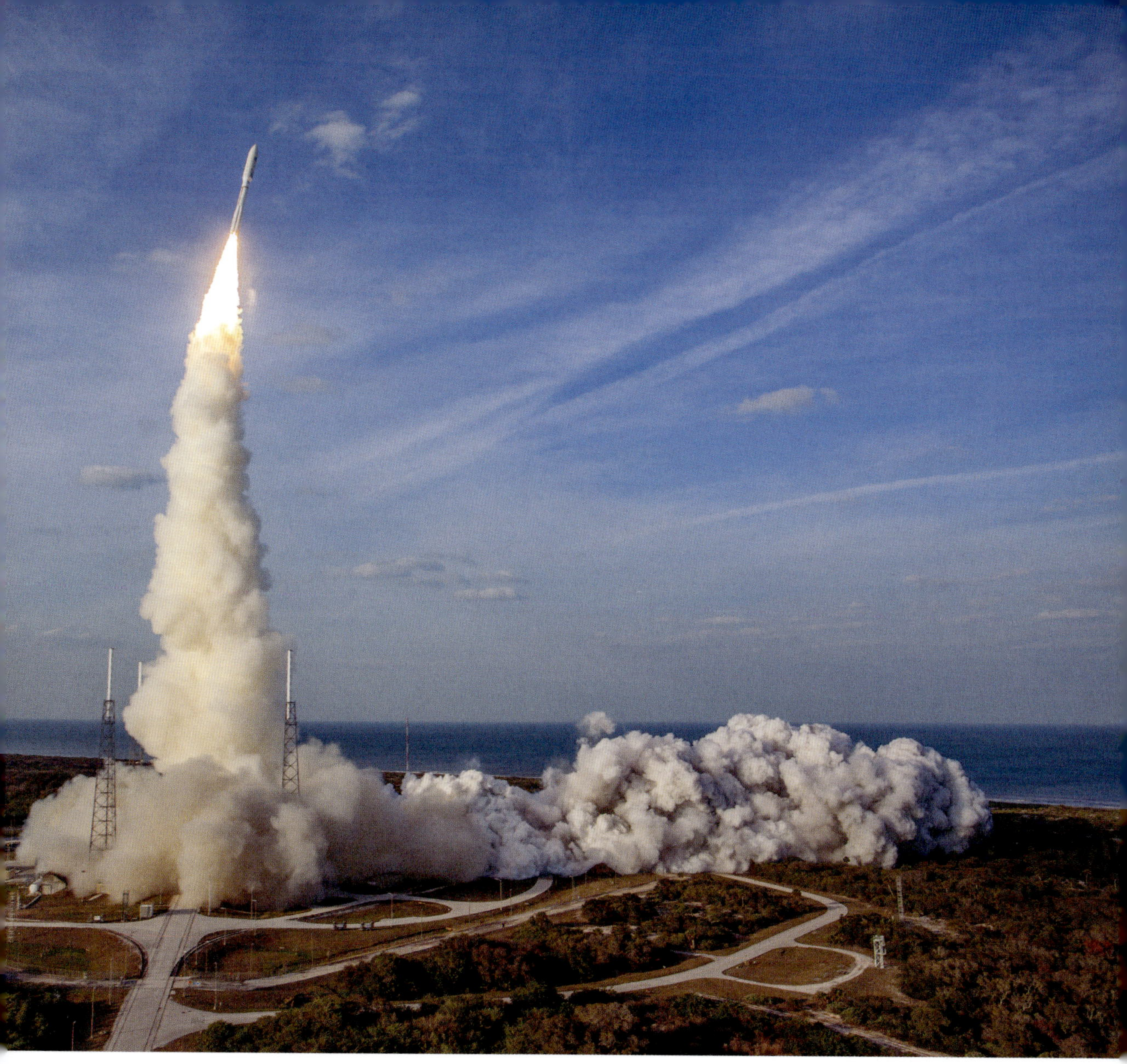

Above the Launch

(previous page, bottom) When possible, I also like to get shots from angles that are above the launch. Being elevated adds a third dimension to the photos, bringing more of the ground into play as a backdrop, as well as sky. This fish-eye view of a Falcon 9 was taken with a camera mounted to one of the four lightning-protection towers at pad 40. The towers themselves can be over 300 to 400 feet in height and must be climbed by ladder.

A Wide View

(above) In this wide-angle shot, captured from a similar height, an Atlas V begins to arc downrange and head out over the Atlantic with its satellite delivery on board. Taken from the top of the assembly building a third of a mile away, a super-wide-angle photo like this can make the rocket seem small. Make no mistake, though, the rocket stands as tall as a twenty-story building in height, and the lightning towers are the same height as in the photo on the bottom of the previous page.

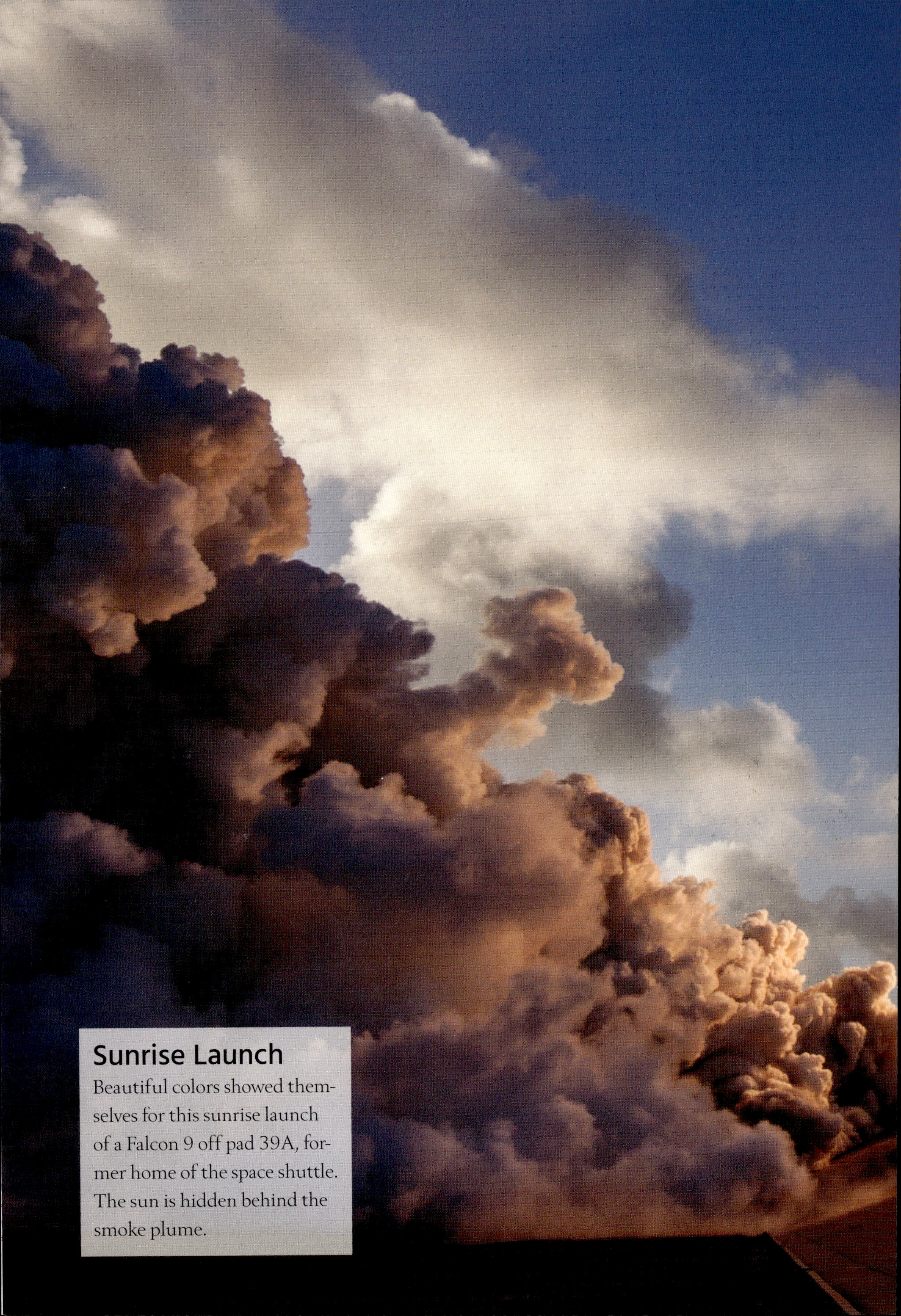

Sunrise Launch

Beautiful colors showed themselves for this sunrise launch of a Falcon 9 off pad 39A, former home of the space shuttle. The sun is hidden behind the smoke plume.

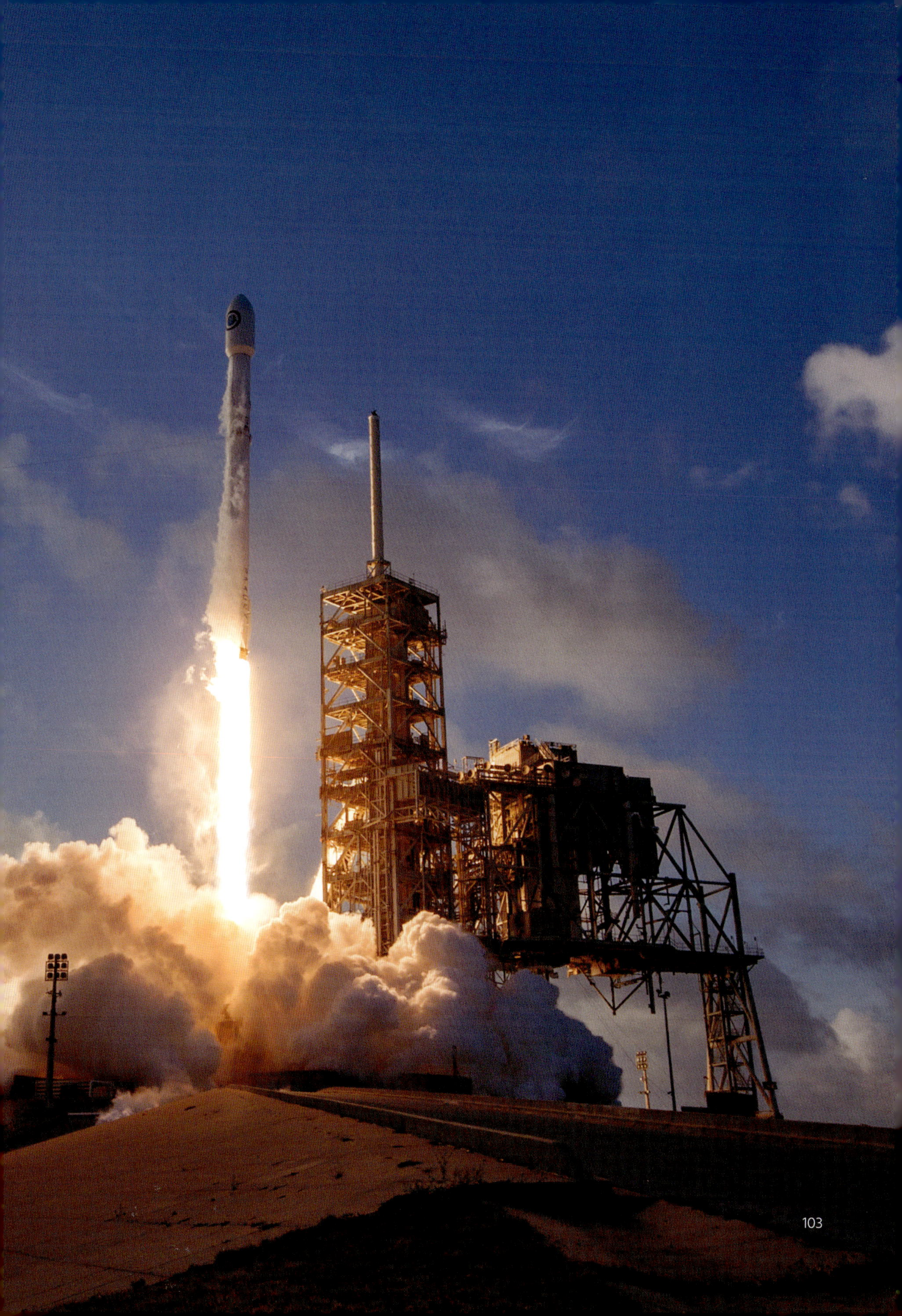

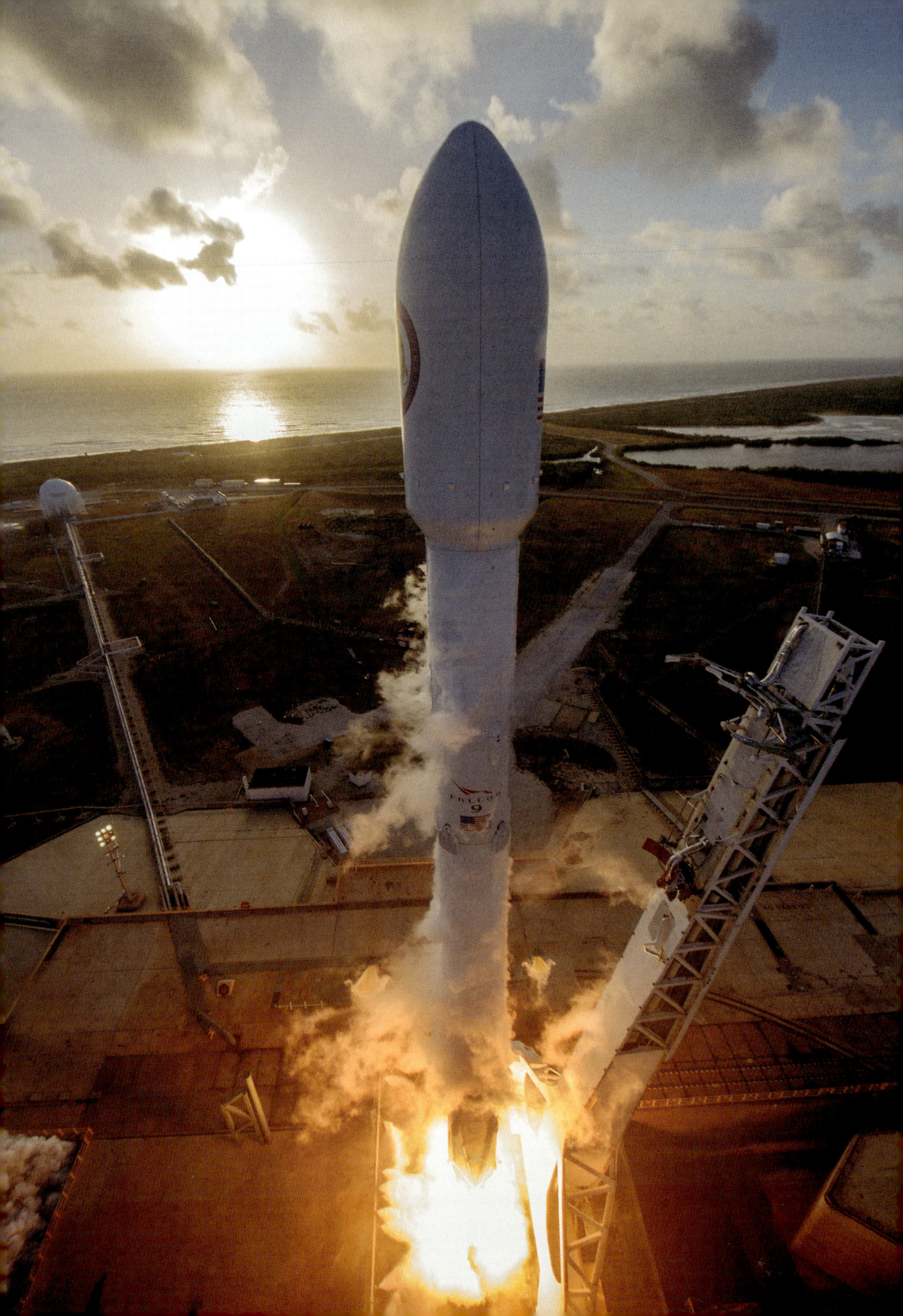

Sunrise Ascent

(previous page) A Falcon 9 begins its sunrise ascent as seen from the very top of the launch tower next to it at pad 39A.

Evening Silhouette

(top) An Atlas V rocket appears in silhouette against the sun in the early evening. For launches when I know the sun will be at a low angle, morning or evening, I always calculate the position it will be at if I want a silhouette like this.

After Sunset

(center) A Delta IV launching after sunset.

Nighttime

(bottom) At night, the golden-white color of the rockets' flames illuminate a Falcon 9's exhaust from behind.

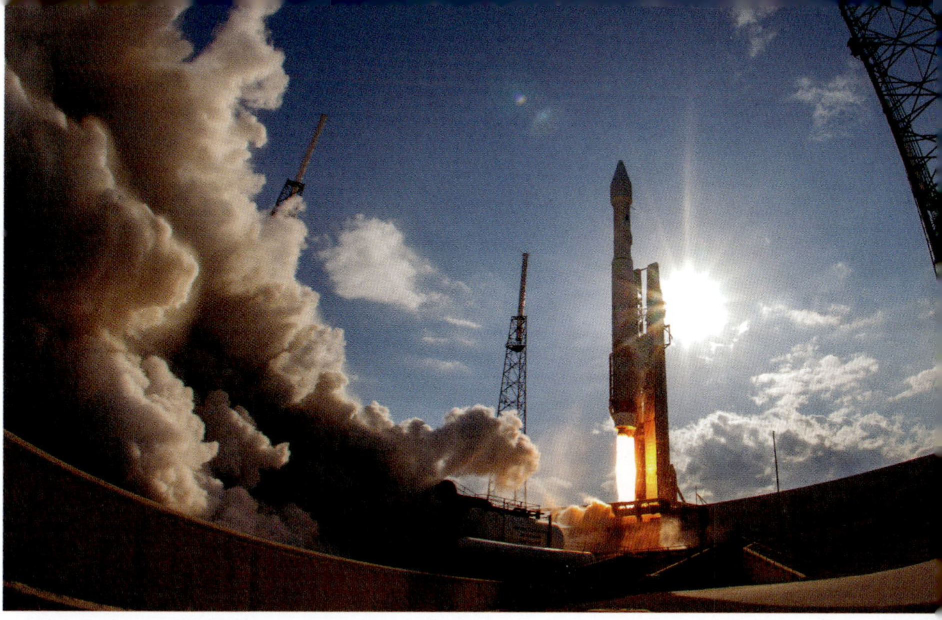

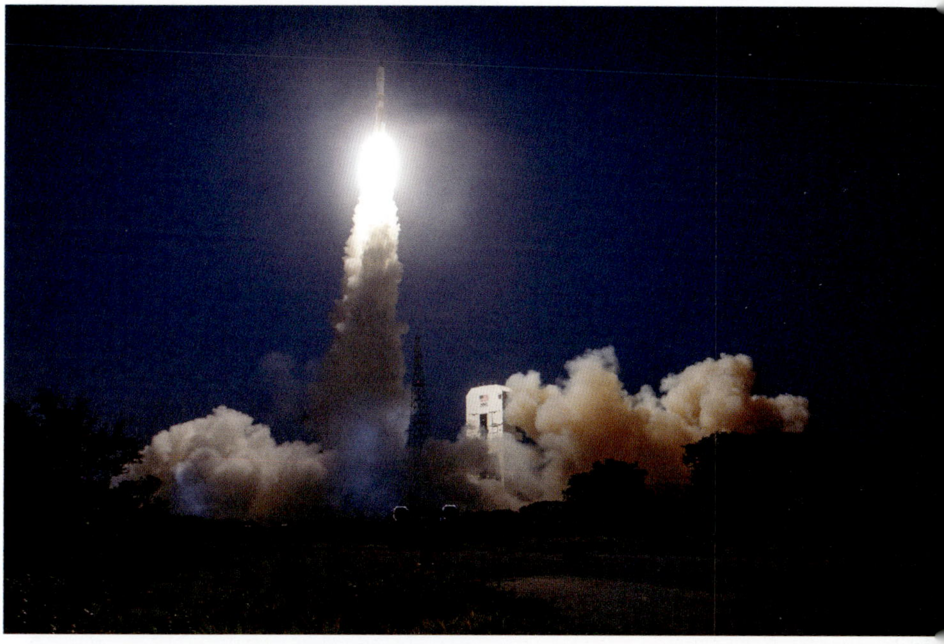

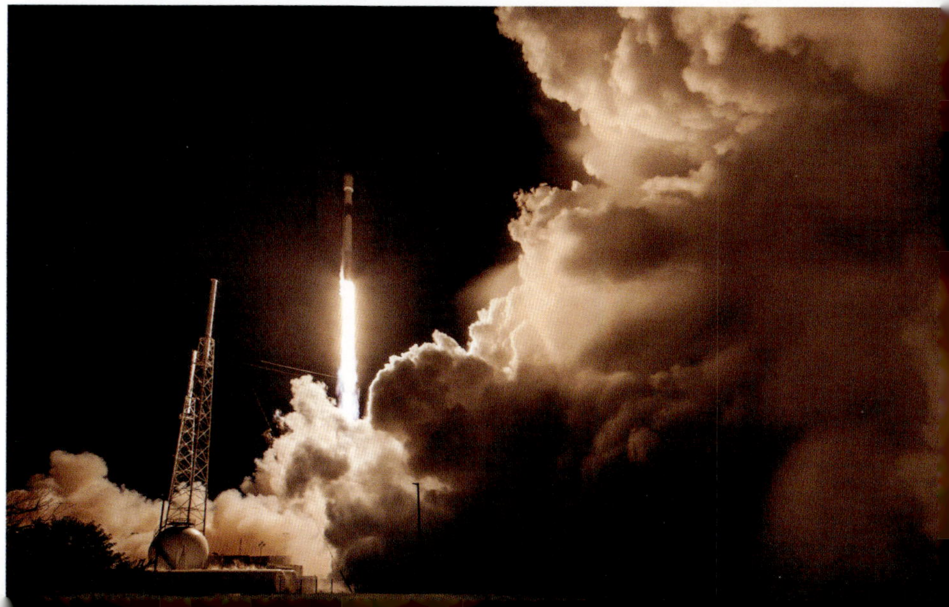

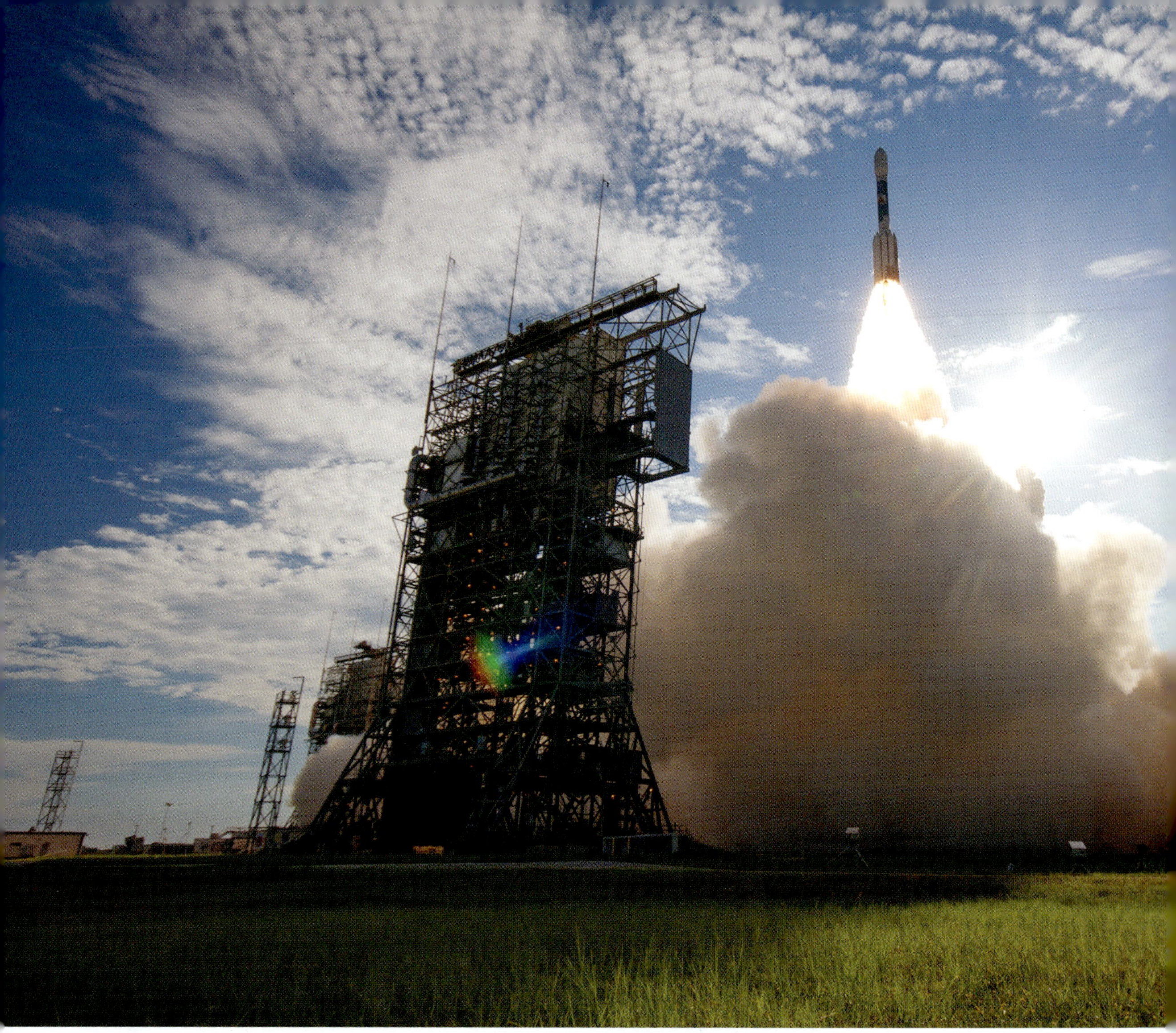

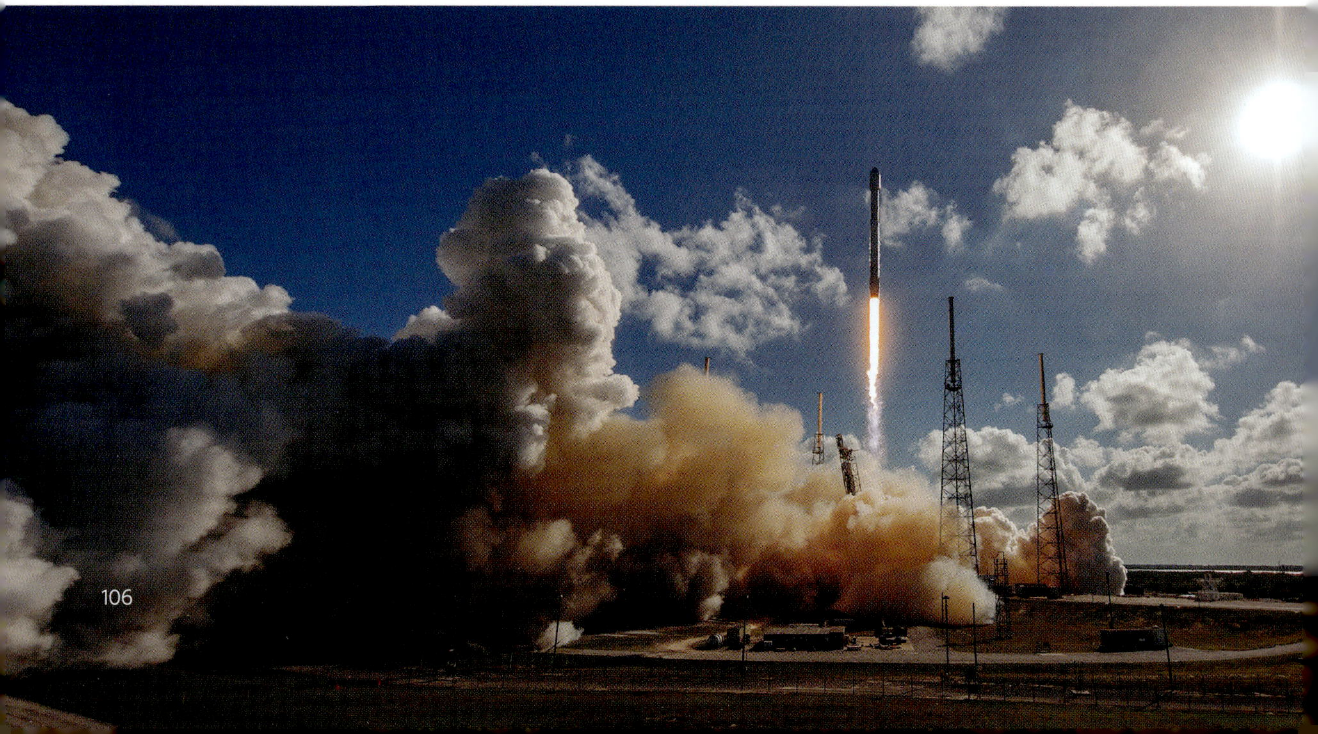

Complex 17

(top left) A Delta II rocket lifts off on a beautiful morning on a mission to the moon. This was the final launch from this historic launch pad, Complex 17 at Cape Canaveral. The pad had been in use since the 1950s. The Delta II rocket itself was retired a few years later as well.

Panorama

(bottom left) A panorama taken from an elevated tower near the launch pad captures this Falcon 9 as it begins a late-afternoon launch.

Maiden Flight

(below) The maiden flight of the Delta IV-Heavy in 2004 was notable, as the fireball that erupts at ignition (a normal part of the launch of this all-hydrogen rocket) scorched the brown foam that lines the outside of the rockets' fuel tanks, turning them black.

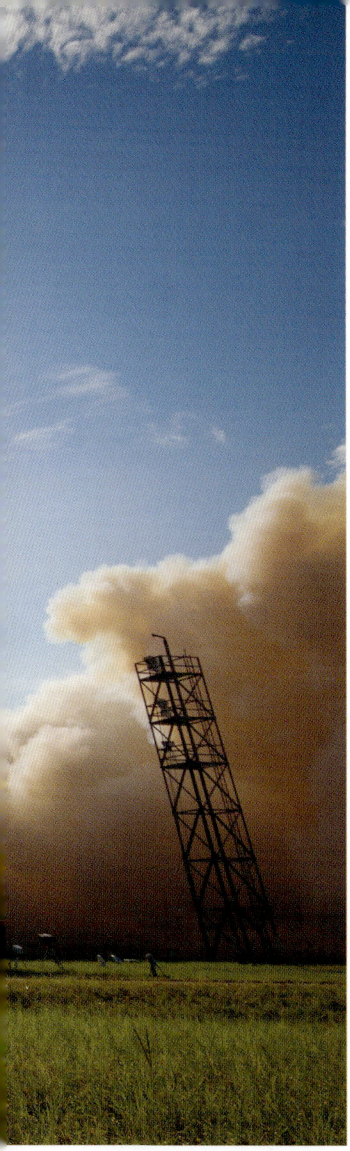

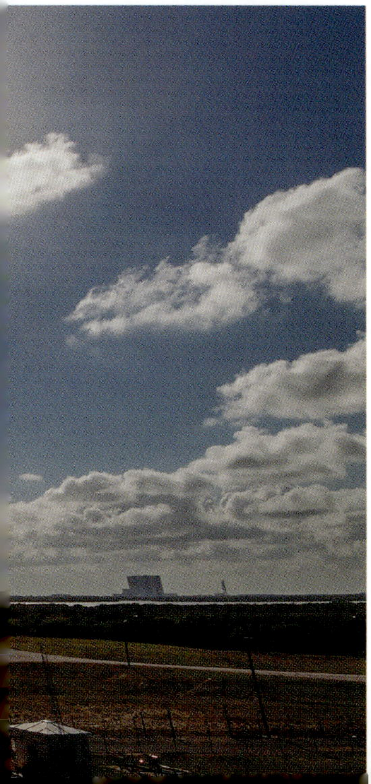

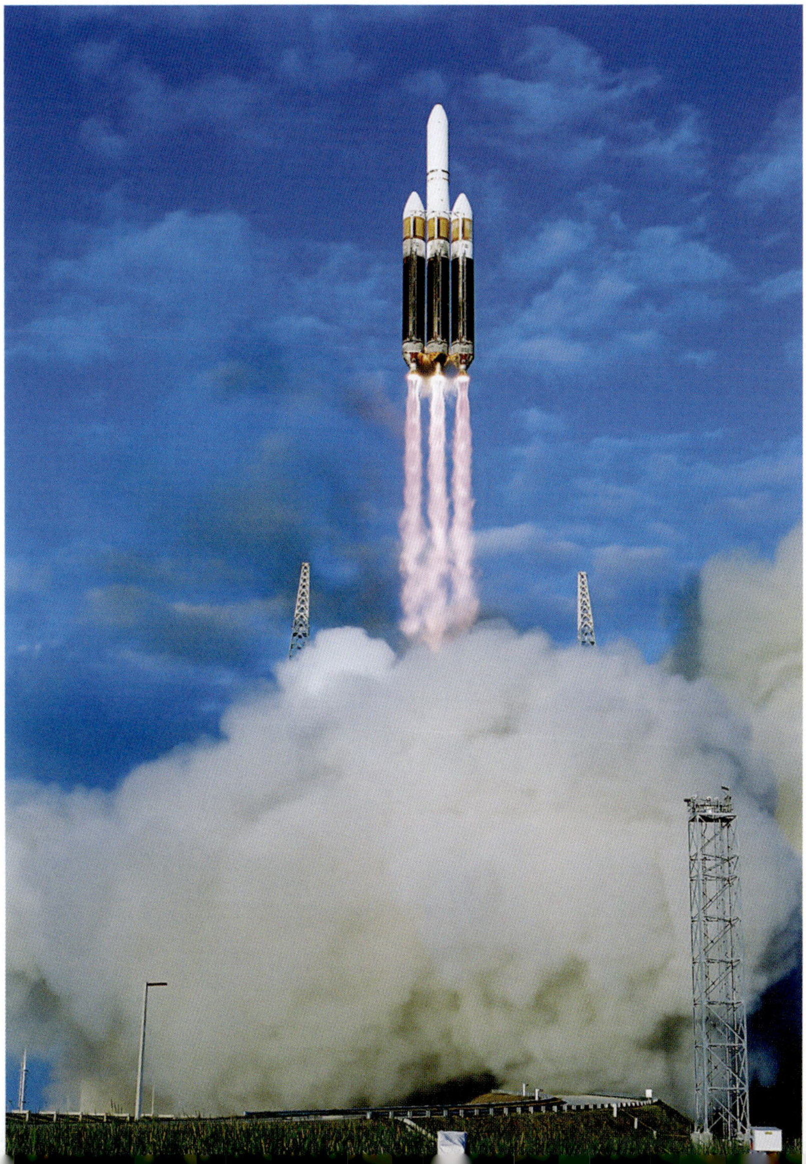

Composition

(top) Careful calculation of where the sun will be in the sky, or aligning the rocket (here, a Falon 9) with the sun the day before launch can help get a shot like this. It can also be pure luck, if the launch is delayed or perhaps you weren't paying attention to it, which was the case here!

A Fleeting Moment

(bottom) As the rocket climbs higher and out of the frame, the last shot in the sequence will frequently be the one before the smoke plume engulfs the camera or blocks the rocket. Sometimes, this means a dramatic, unexpected shot, such as this Delta IV launch.

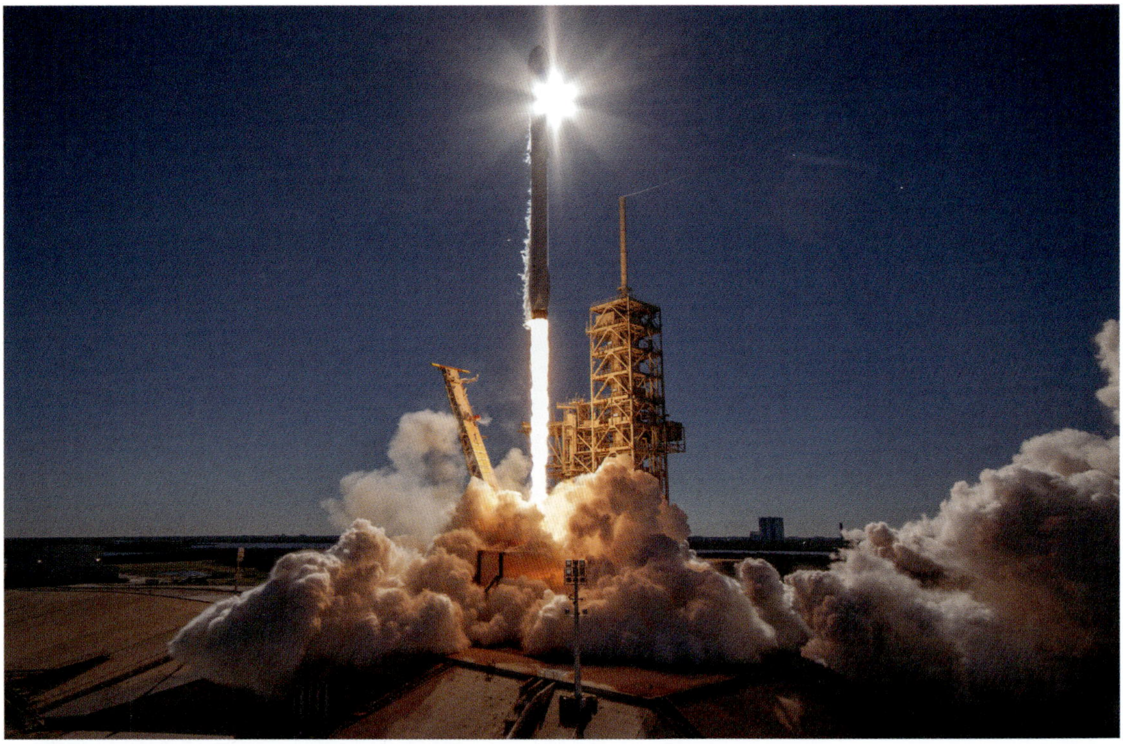

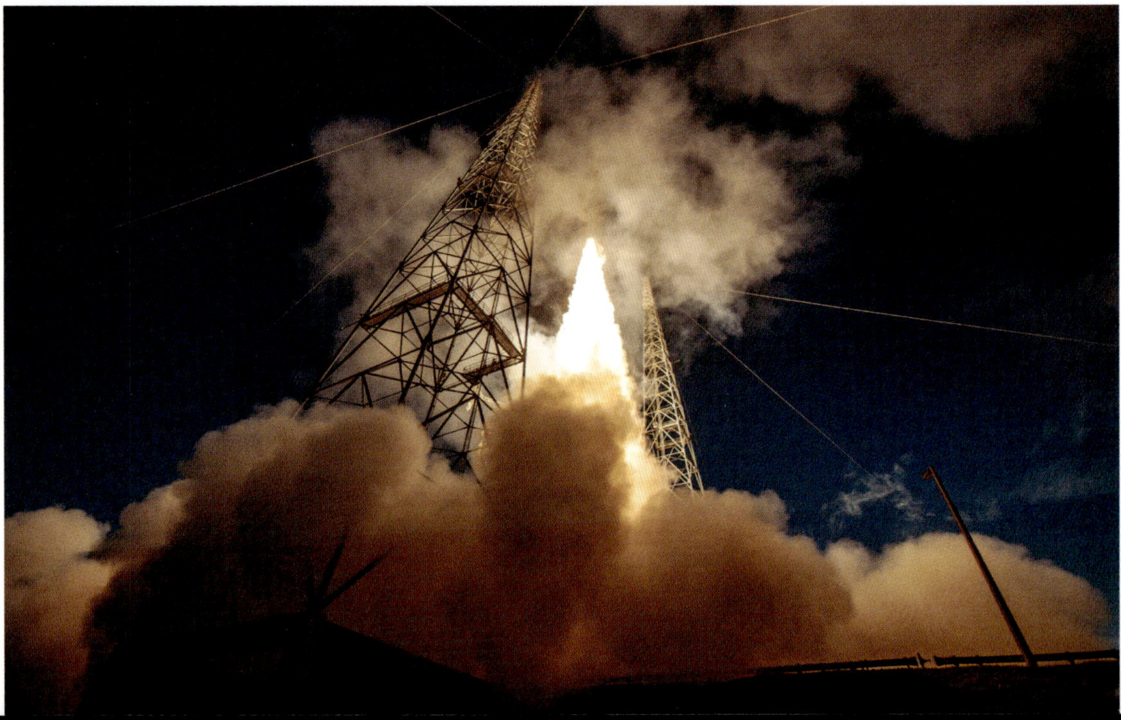

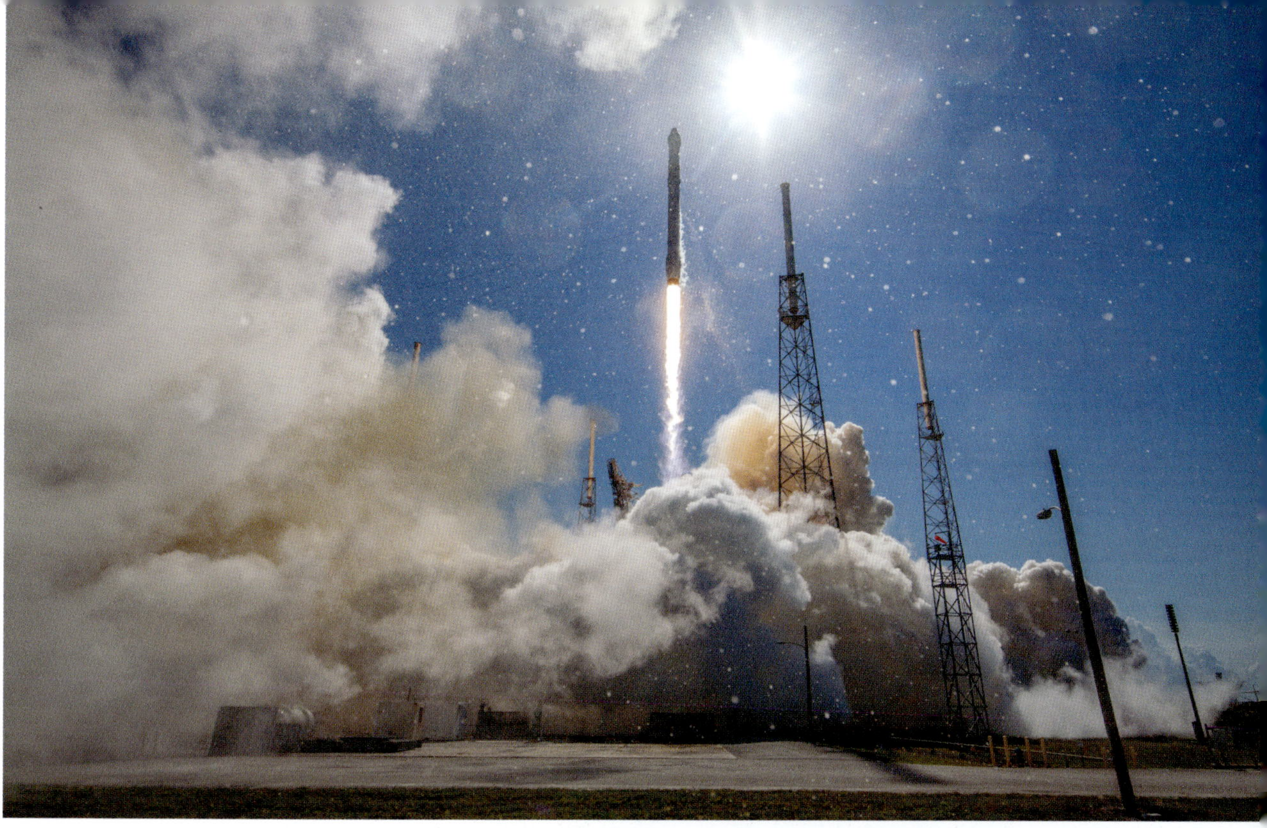

Water Spray

(above) In this unique shot, what appears to be snowflakes are in fact droplets of water spray being flung at the camera, nicely reflecting the sunlight behind them. The photo before this one was clear; in the image after, the lens was coated and no longer able to produce a clear shot of the Falcon 9.

Atlas V

(right) Reeds frame this Atlas V rocket as it ascends from pad 41 at the Cape in 2013.

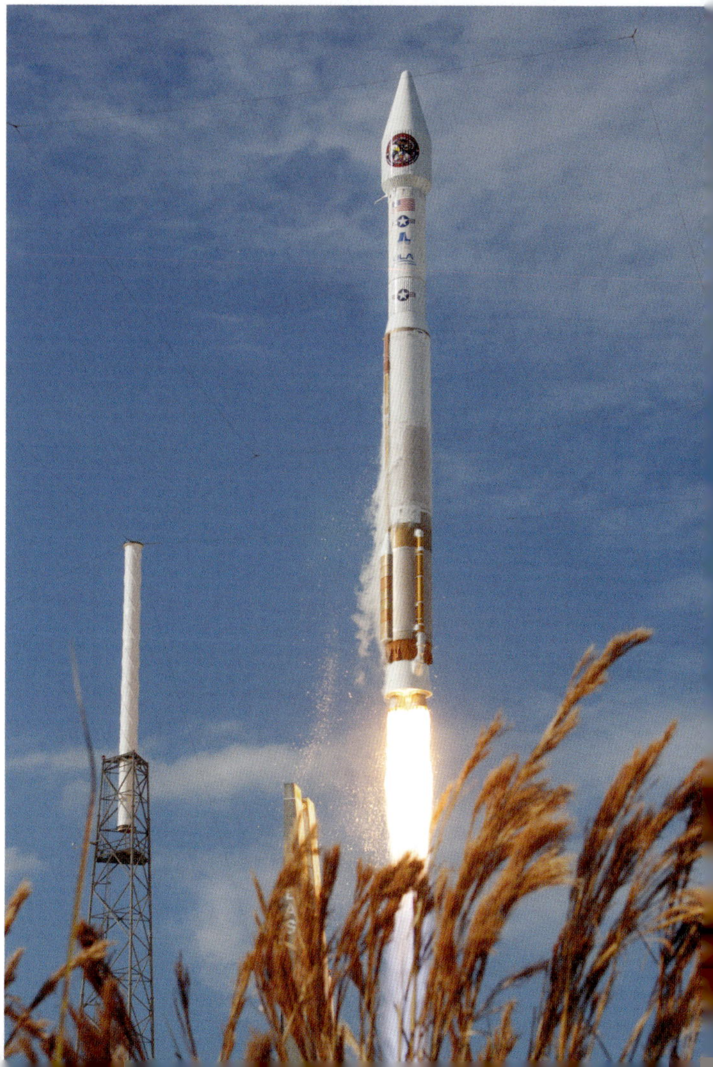

A Super-Wide Angle

(below left) A super-wide-angle lens captured this Delta IV rocket long enough to see it arcing over well away from the launch pad, as it begins its trajectory over the Atlantic Ocean and into space.

Minotaur I

(below right) A Minotaur I launching at sunrise from Wallops Island, Virginia. The smoke trail can make all the difference and complete the composition in photos such as these.

Out of the Fog

(following page) Vandenberg Air Force Base, situated along California's spectacular Pacific Coast, can be especially pretty. Framed by yellow flowers, a United Launch Alliance Delta IV rises out of the mist at Vandenberg Air Force Base, California, on its way out over the Pacific Ocean. Fog is a common occurrence at Vandenberg, and clear launches like this occur perhaps 50 percent of the time.

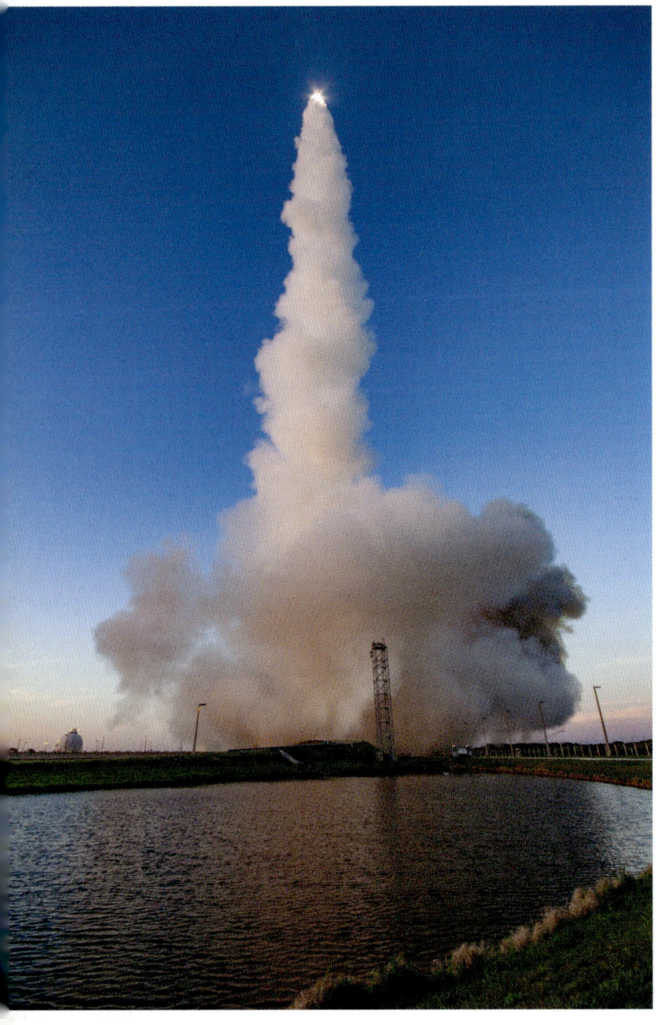

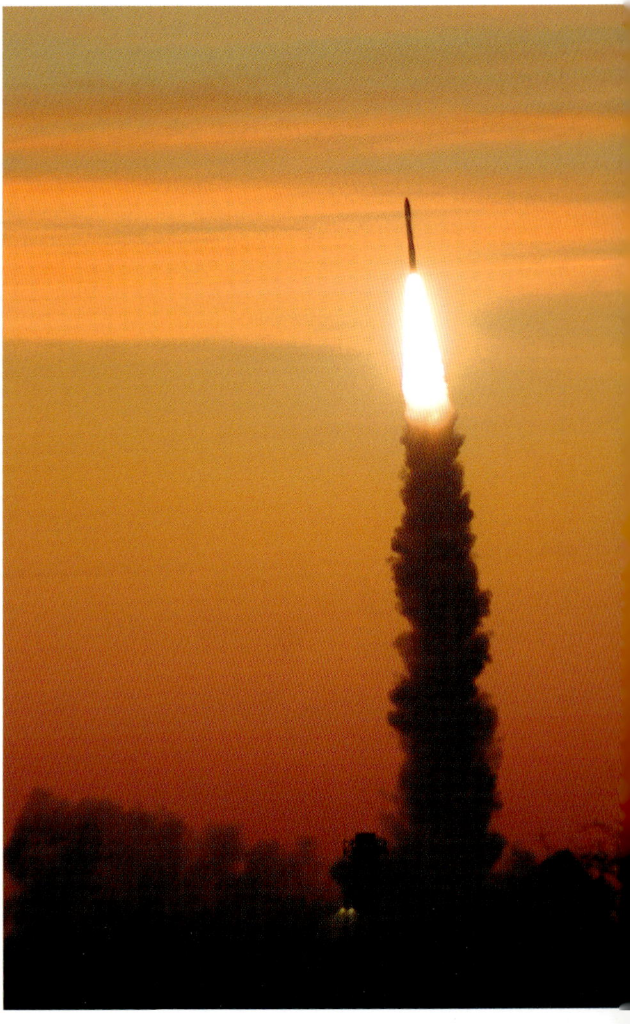

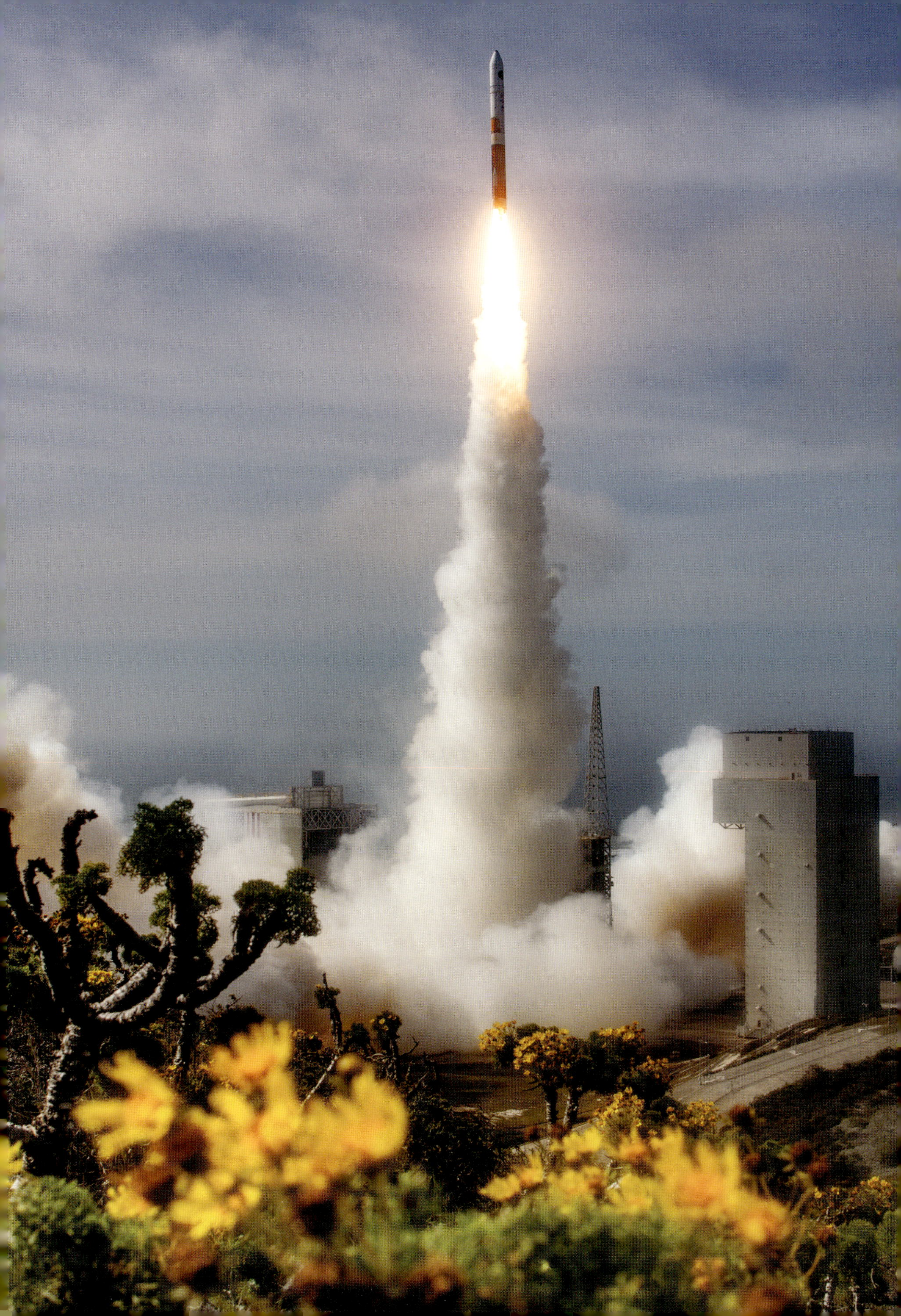

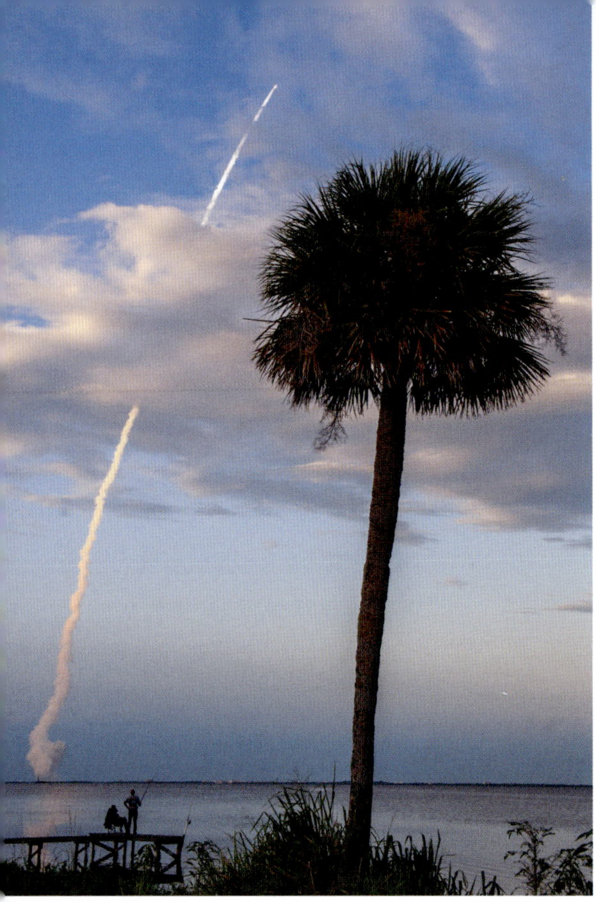

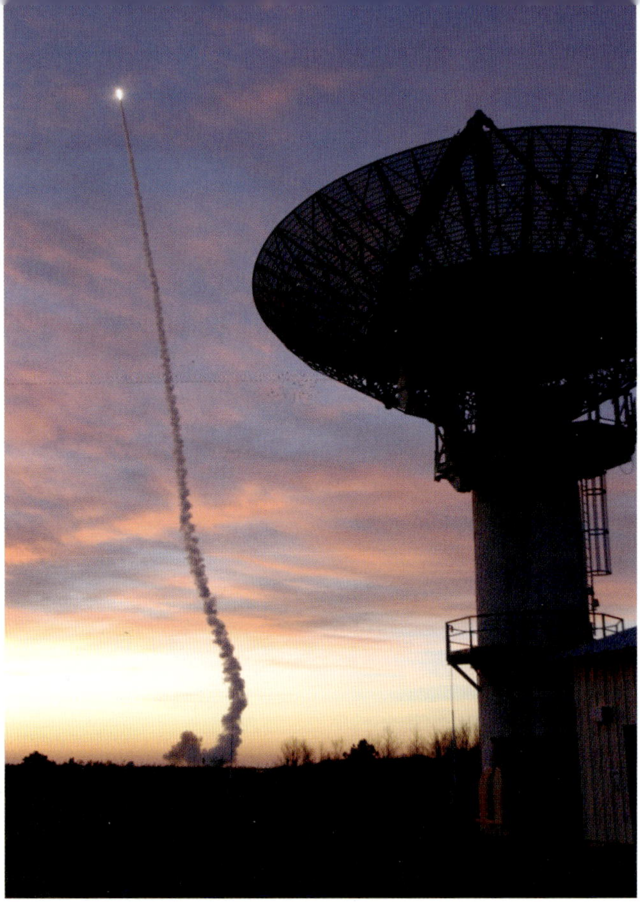

More Smoke Trails

(above left and right) Two more examples of why a launch with a smoke trail can be more photogenic when it leaves the pad and climb high: On the left, a fisherman watches a Delta IV rocket zoom into orbit at sunset at Cape Canaveral. On the right, a Minotaur I rocket speeds away from Wallops Island, Virginia, just before sunrise.

Memorable Sight

(below) During rare pre-dawn launches from Florida (or post-sunset from California), light coming from the sun still well below the horizon can scatter off ice crystals that form from the rocket's exhaust (here, from a Falcon 9) when it reaches the cold upper atmosphere. These twilight shows, while rare, have been among the most memorable and can be seen from hundreds of miles away.

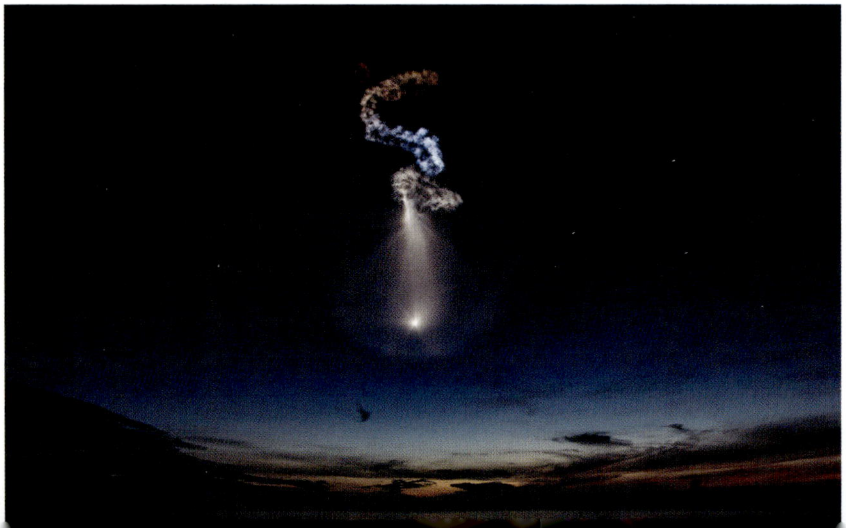

Through the Clouds

(top) As the rocket climbs high into the sky, the opportunity to capture it racing through a cloud, as with the Falcon 9 rocket here. As the rocket climbs high into the sky, the opportunity to capture it racing through a cloud, as seen here, is another sight to aim for.

Max Q

(bottom) The Ares I rocket seen here with a shockwave forming only flew one time before its program was partially canceled. These shockwaves form mainly near the region of maximum aerodynamic pressure, known as Max Q, and last only a second or two at most. Capturing the moment of maximum shockwave can be challenging and unpredictable.

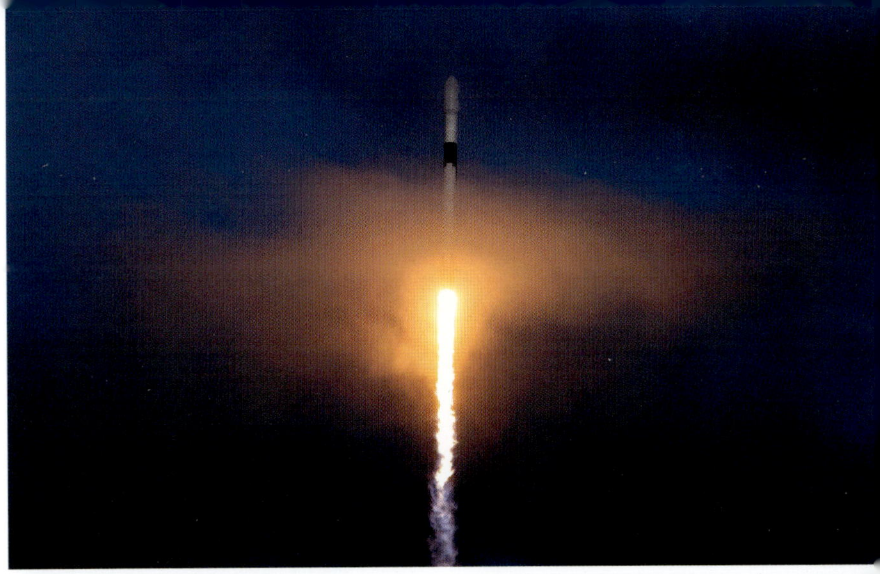

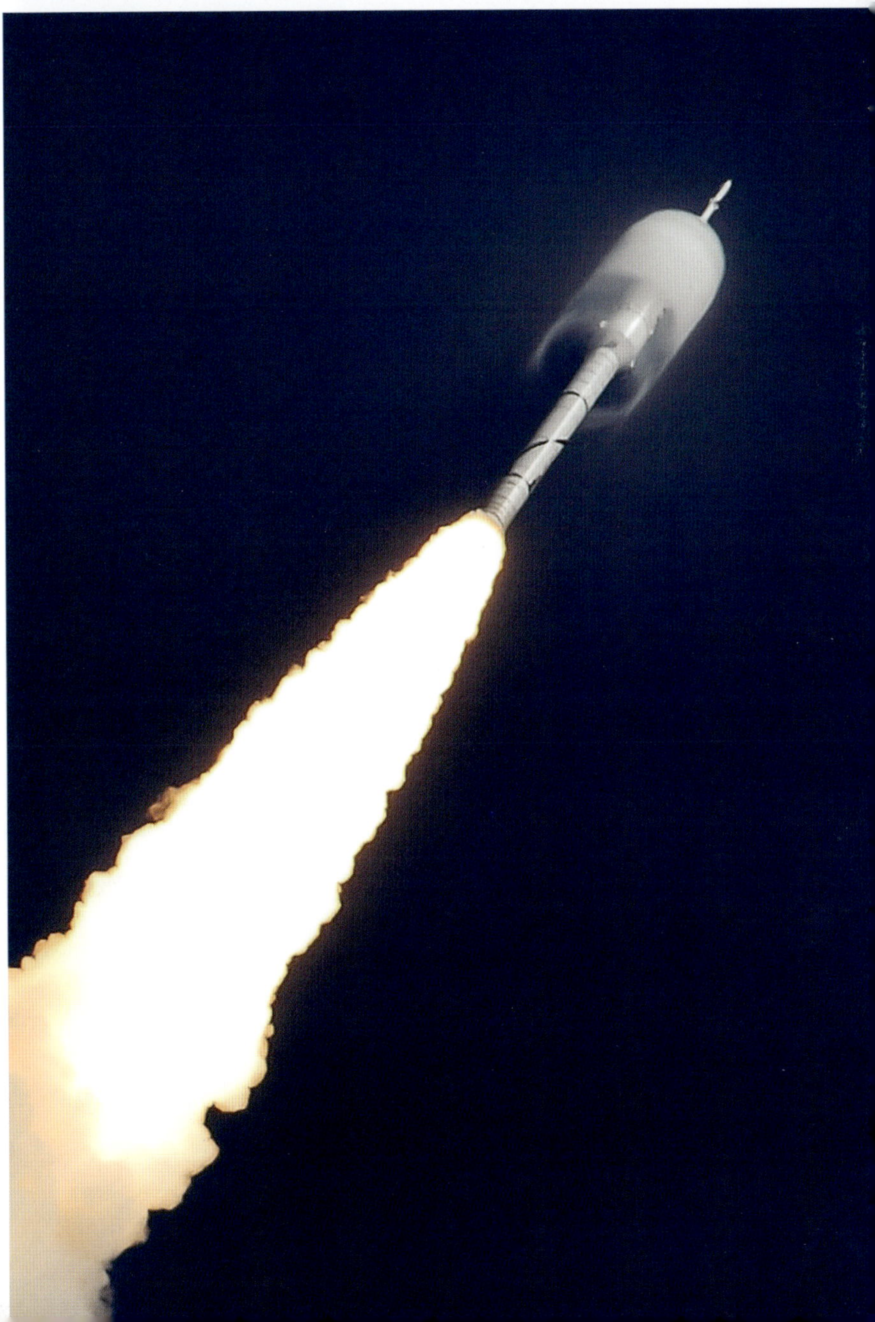

In Action

(right) Photographers, including myself *(center),* Stephen Clark *(right),* and Steven Young *(left)* of Spaceflight Now capture the launch of a Delta II rocket in 2009 from the beach just outside the space center. For some smoke-trail launches, it's nice to be farther away and capture the beauty of the launch and surrounding area in a wide-angle photo such as this.

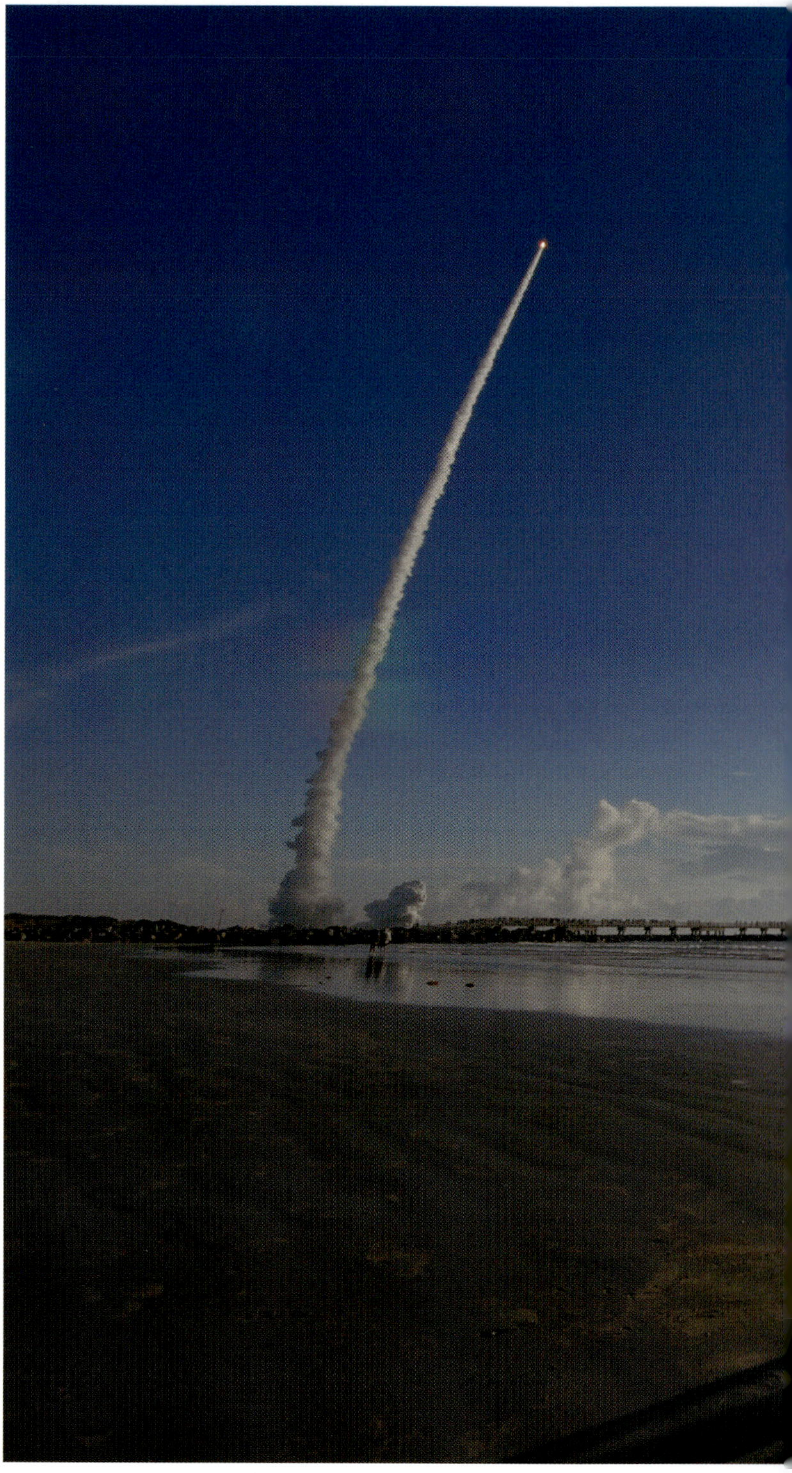

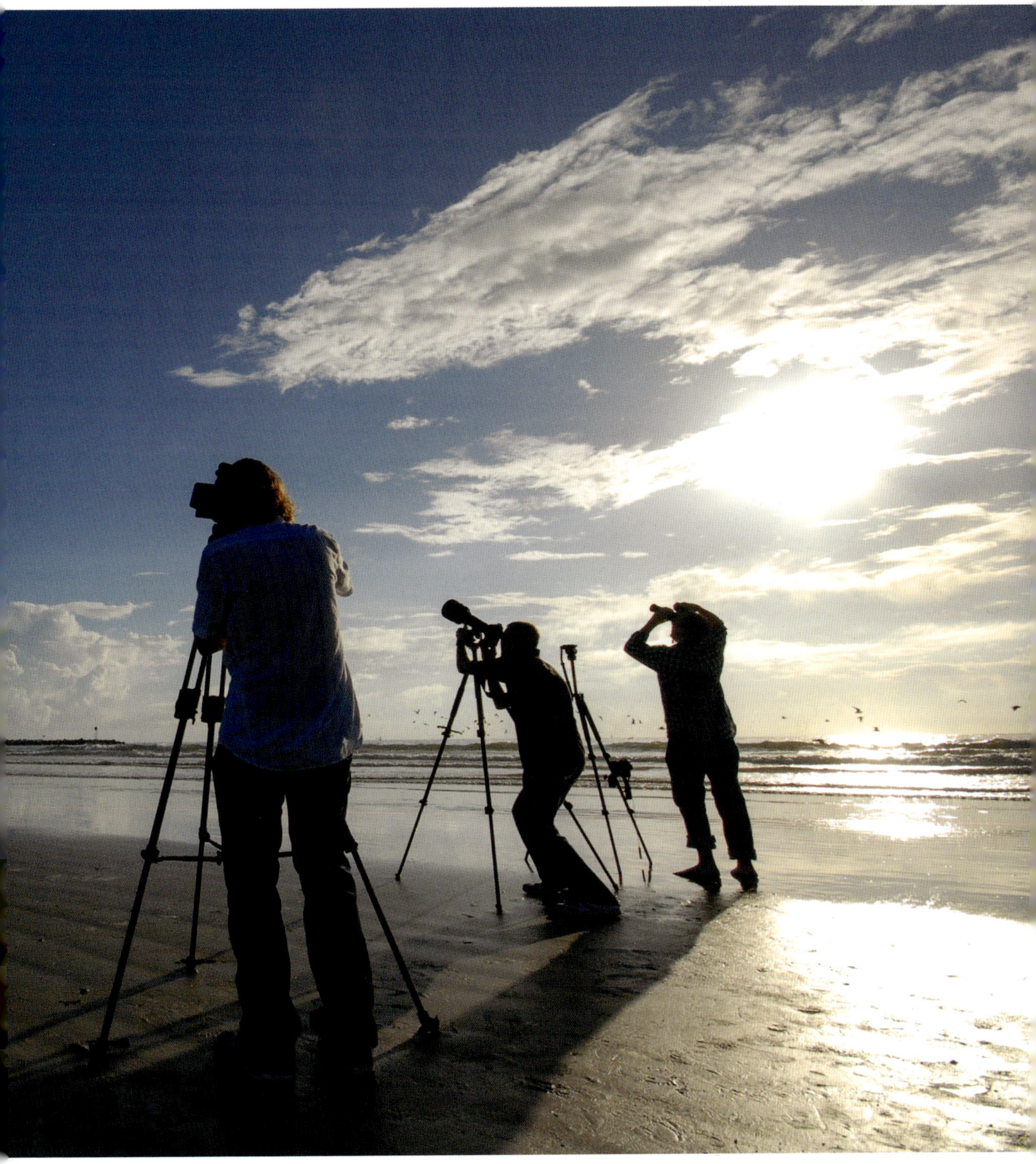

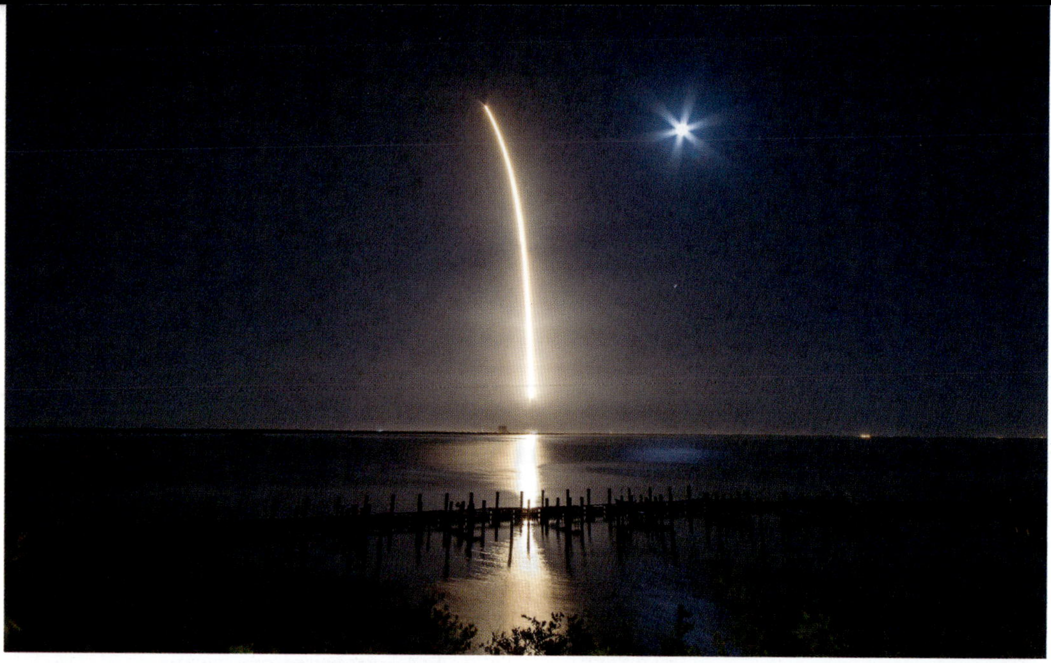

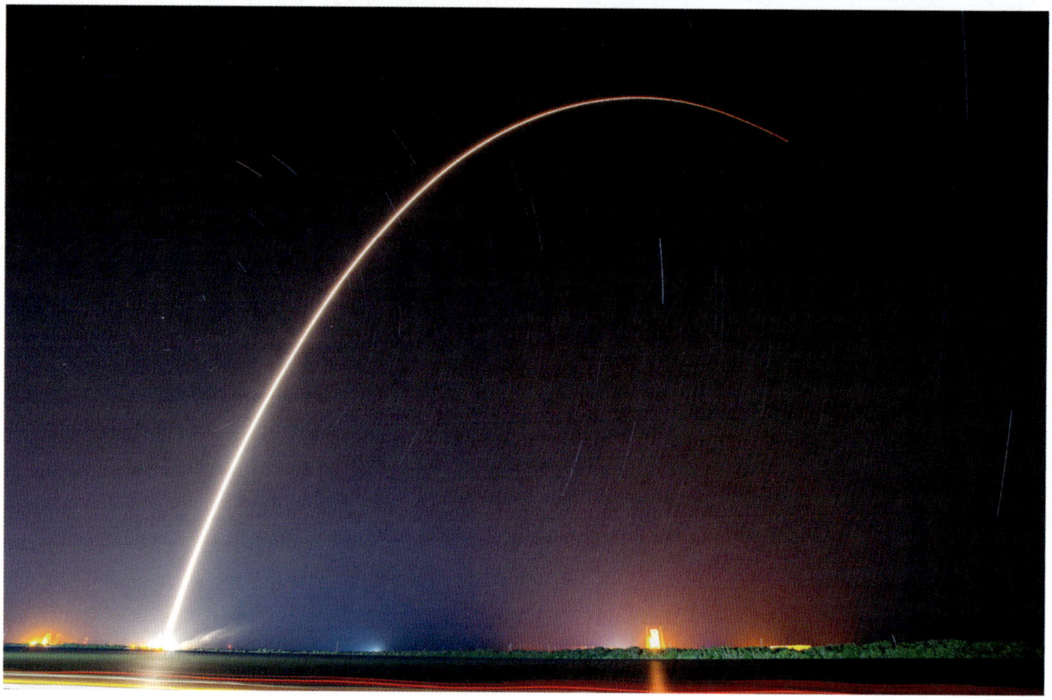

Streak Shots

(this page and following page) Views of a Falcon 9 (above left and right) and an Atlas V. When it comes to night launches, my favorite type of photo is the streak shot. Call it a long exposure of the launch or an arc; it's simply a matter of leaving the shutter open and stopping the lens aperture down so as not to overwhelm the photo with light (these rockets are extremely bright). The key to getting a great streak shot is the foreground, or background in the case of incorporating the night sky into the frame. Streak shots have proved to be among my most popular photos.

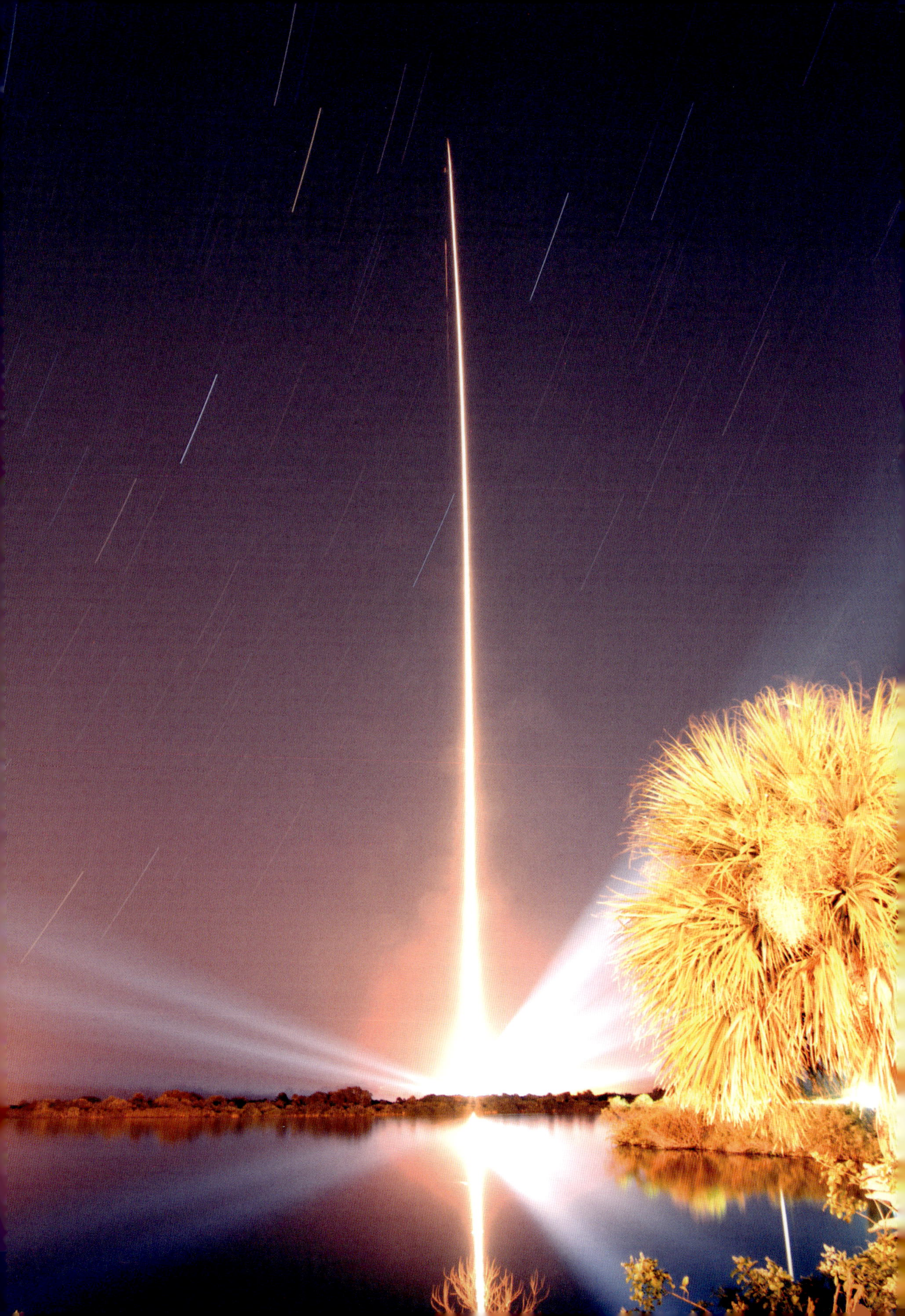

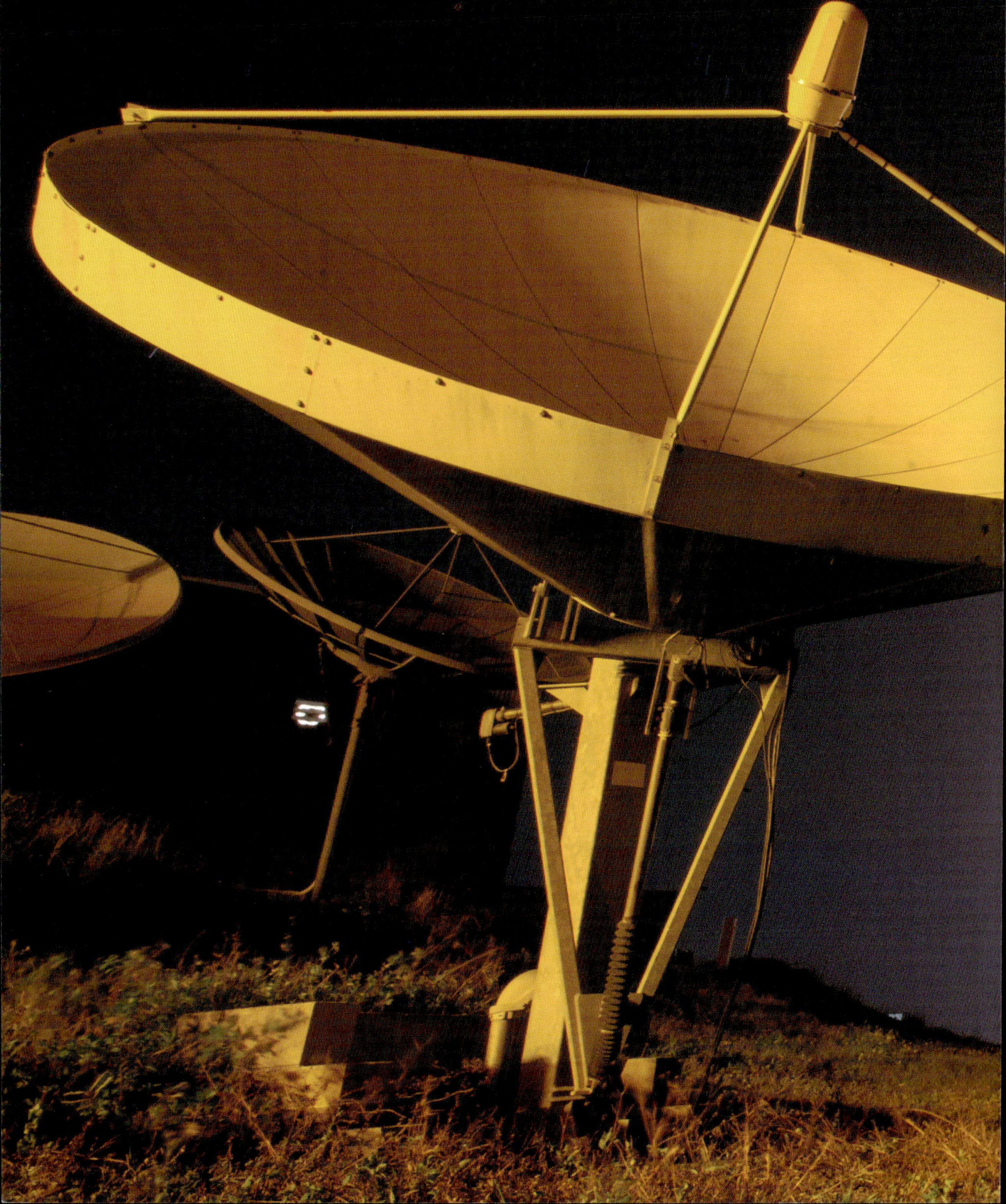

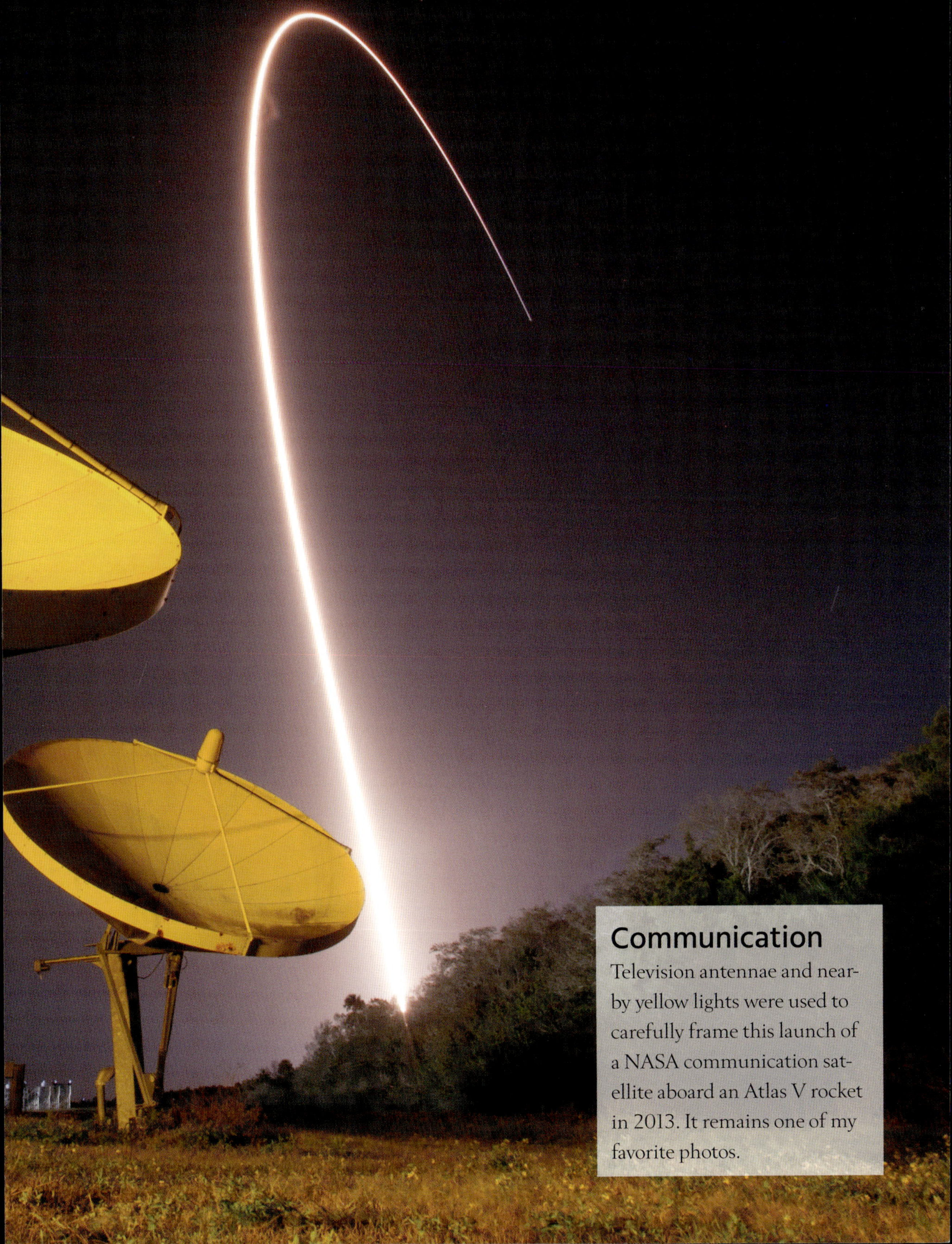

Communication

Television antennae and nearby yellow lights were used to carefully frame this launch of a NASA communication satellite aboard an Atlas V rocket in 2013. It remains one of my favorite photos.

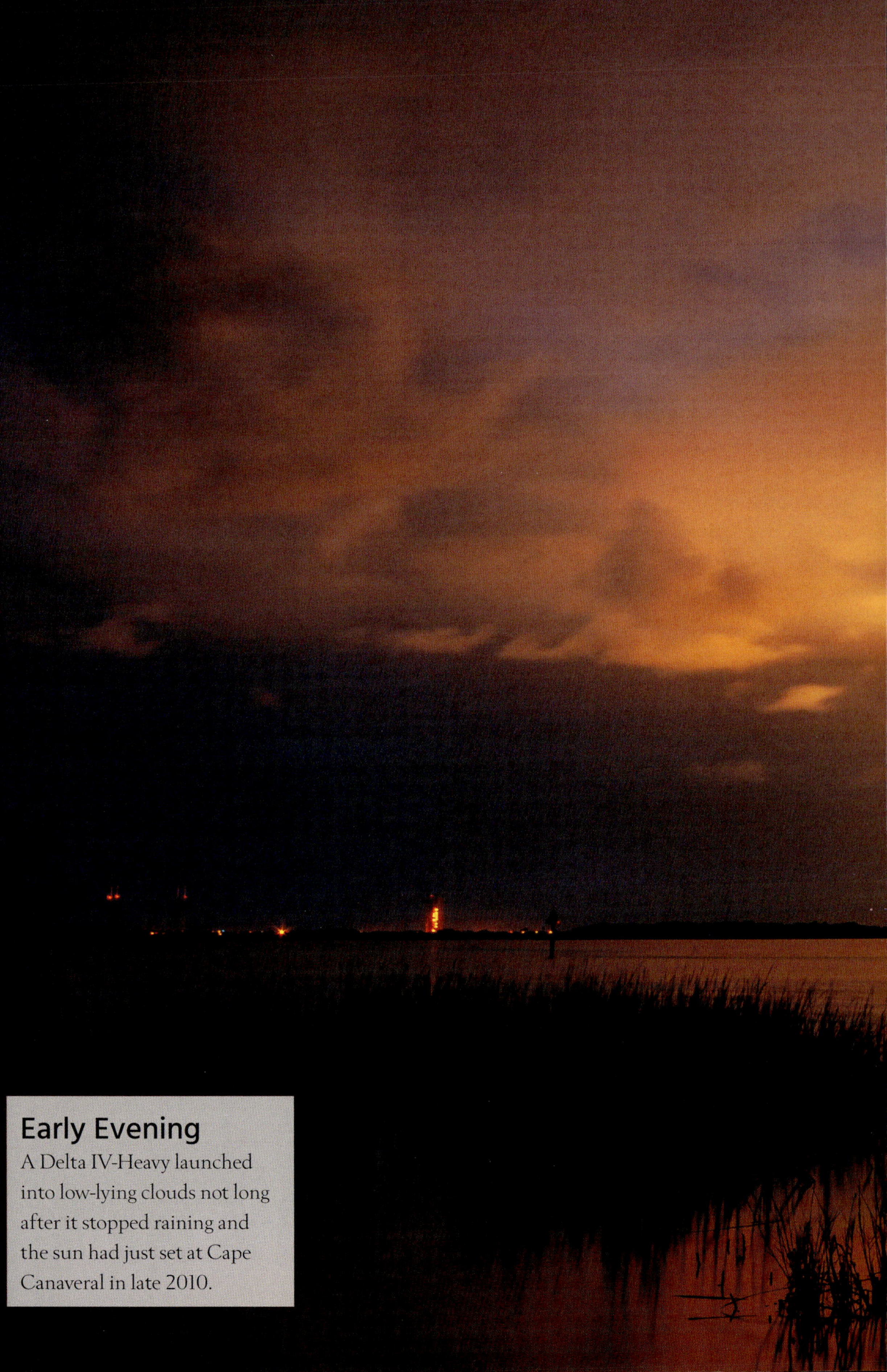

Early Evening

A Delta IV-Heavy launched into low-lying clouds not long after it stopped raining and the sun had just set at Cape Canaveral in late 2010.

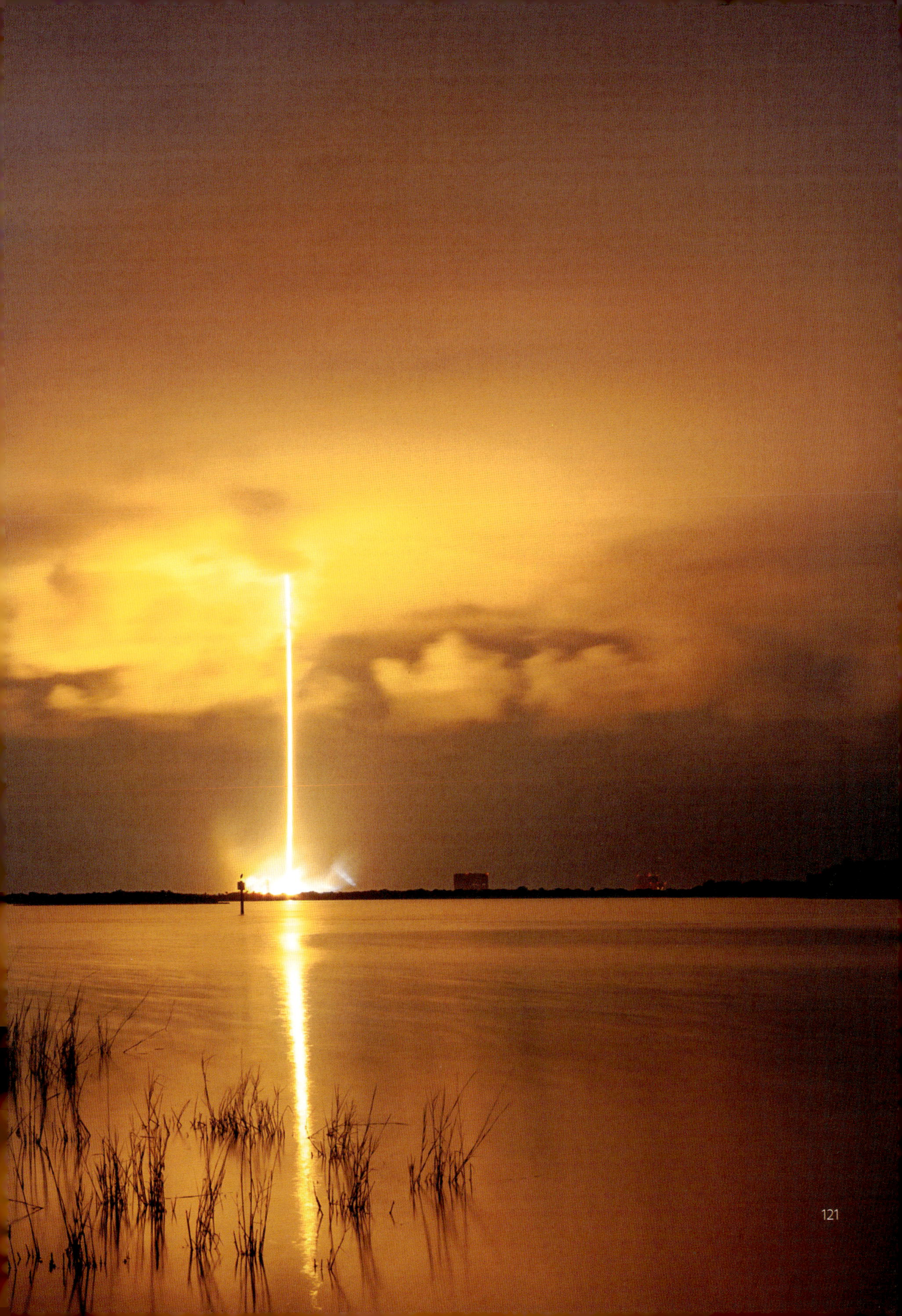

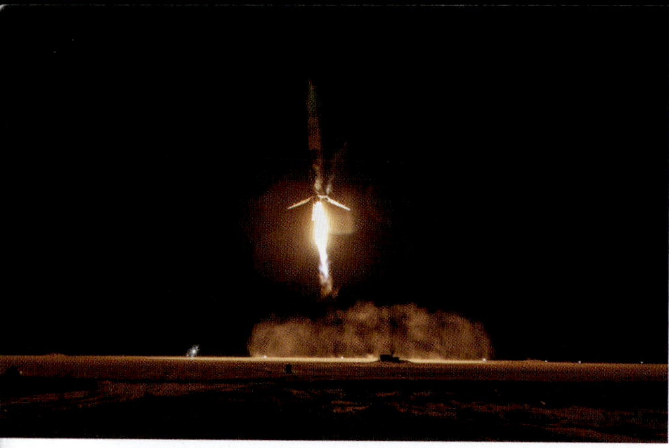

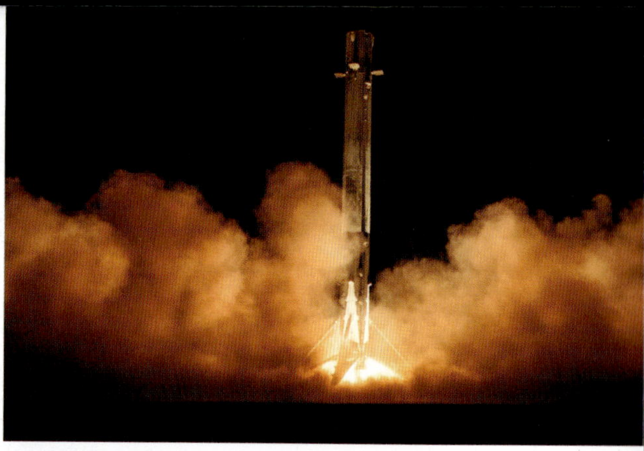

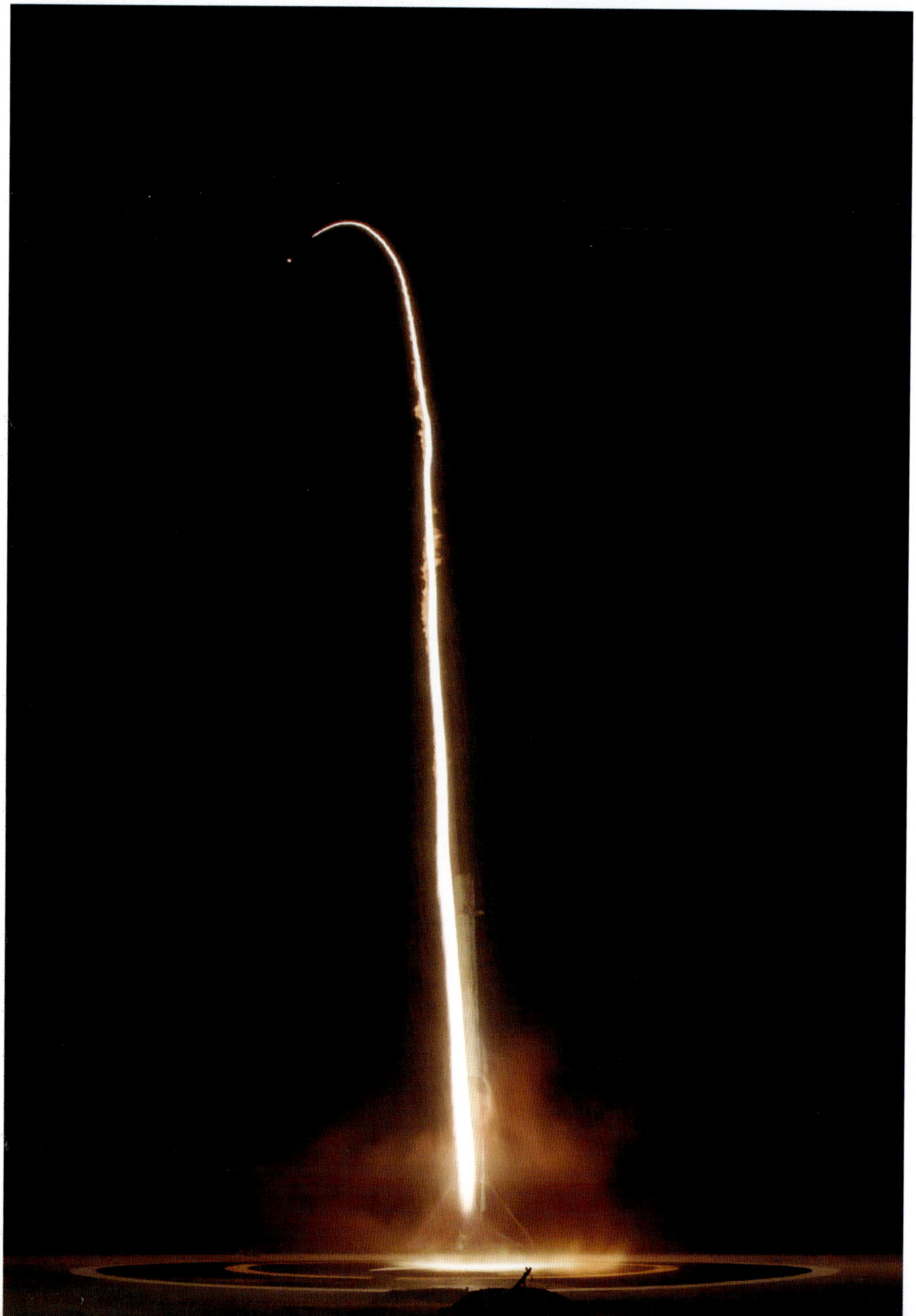

Reuse

(previous page and this page) Finally, what you see on these pages are not rocket launches, though you might not know unless told. In 2015, the space launch company SpaceX achieved their goal of returning the Falcon 9 that just launched back to land with the intention of reusing it. This is the first time an unmanned expendable-style rocket has been made reusable. About three minutes after launching, the first stage of the rocket, like most, separates from the upper stage and payload, which continue on their way. The first stage then ignites its engines a second time, and can either reverse course, landing near the launch pad, or guide it back down toward the ocean, landing on a floating platform at sea, where it is then towed back to land.

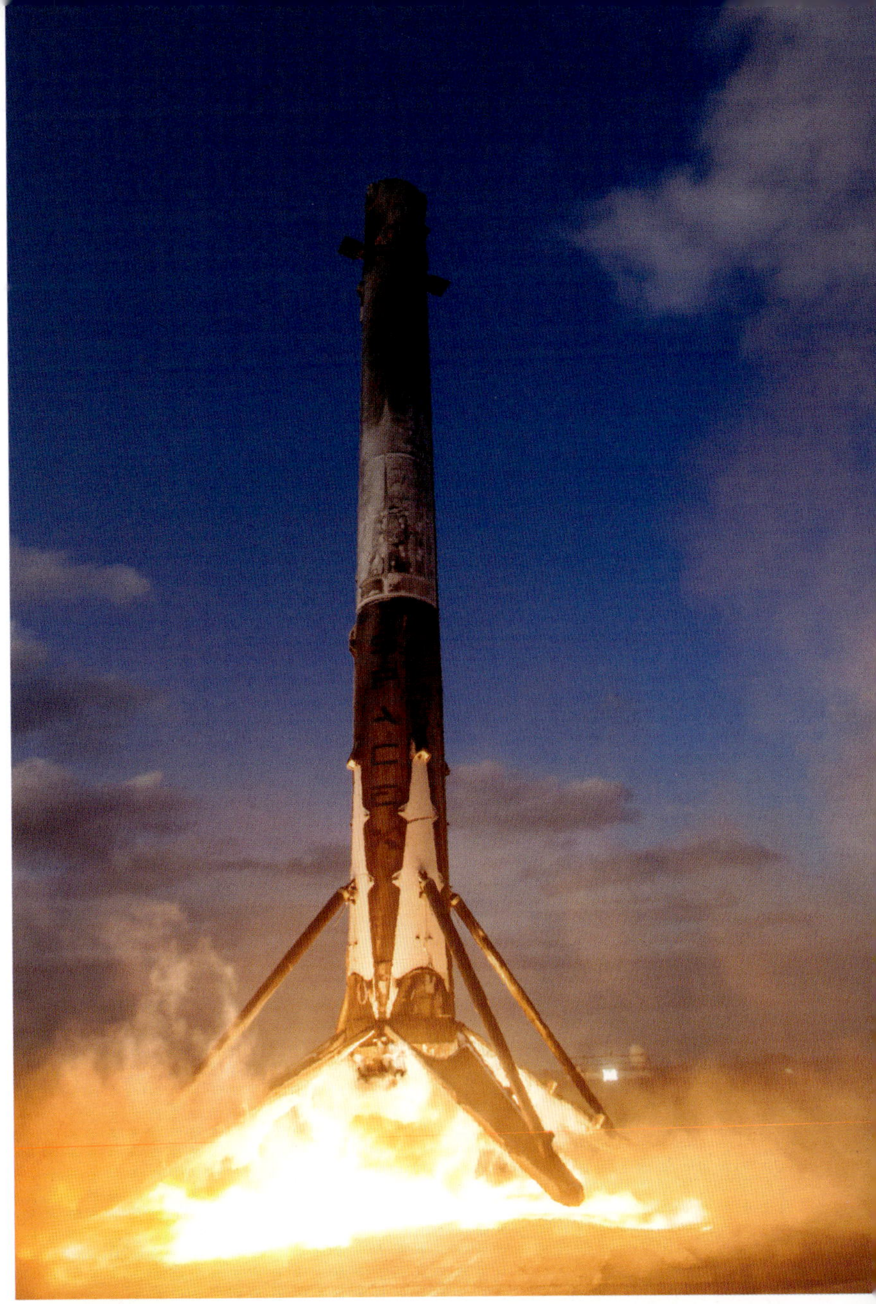

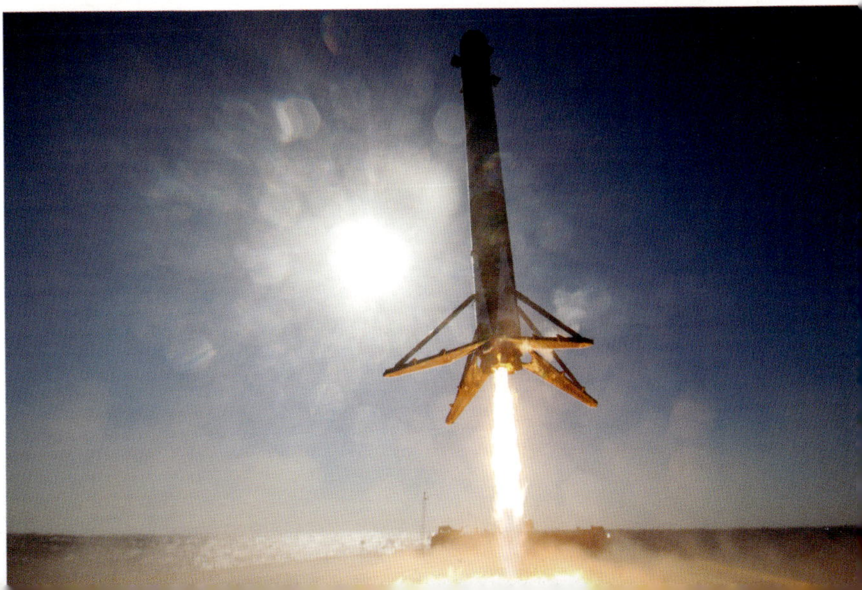

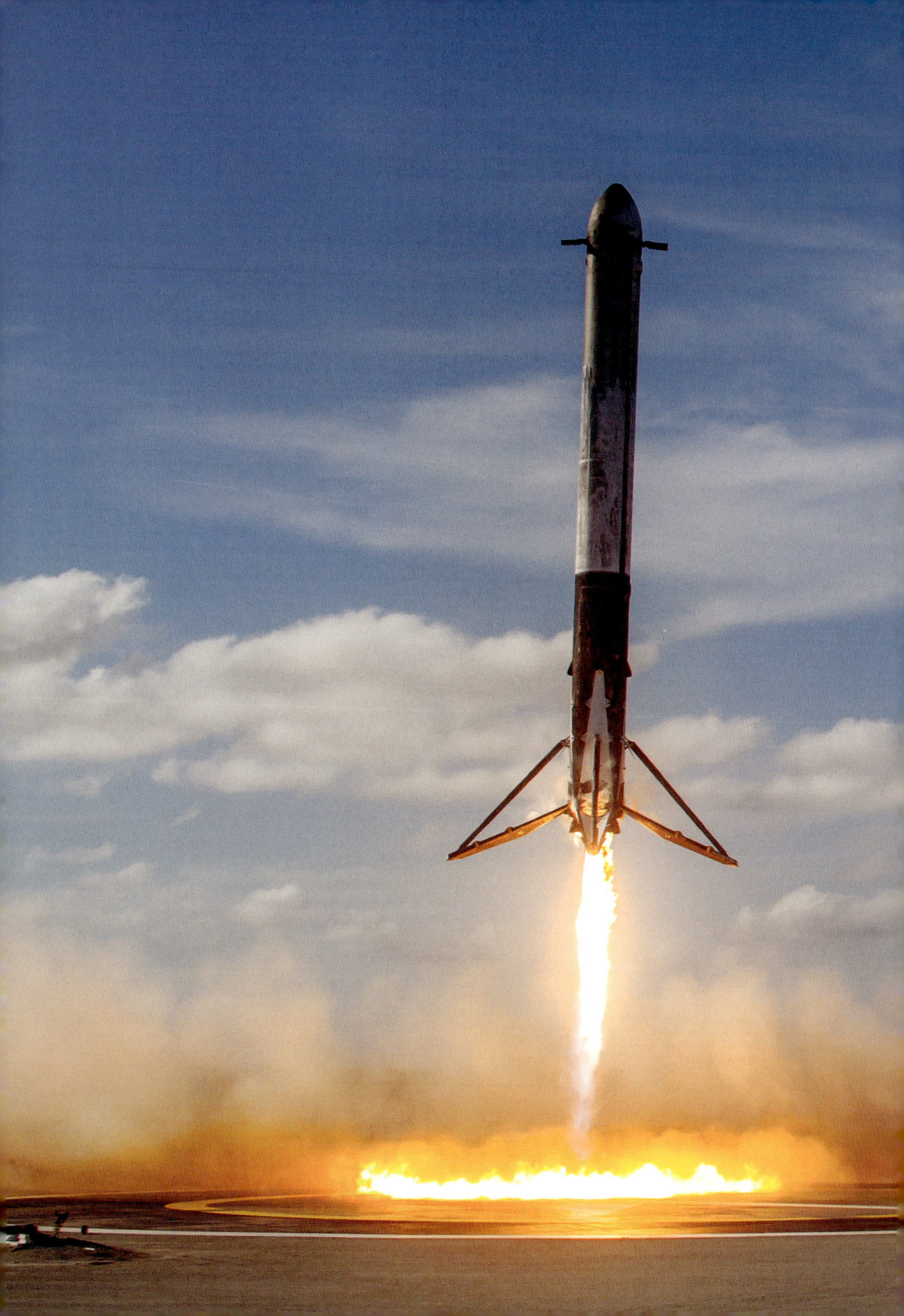

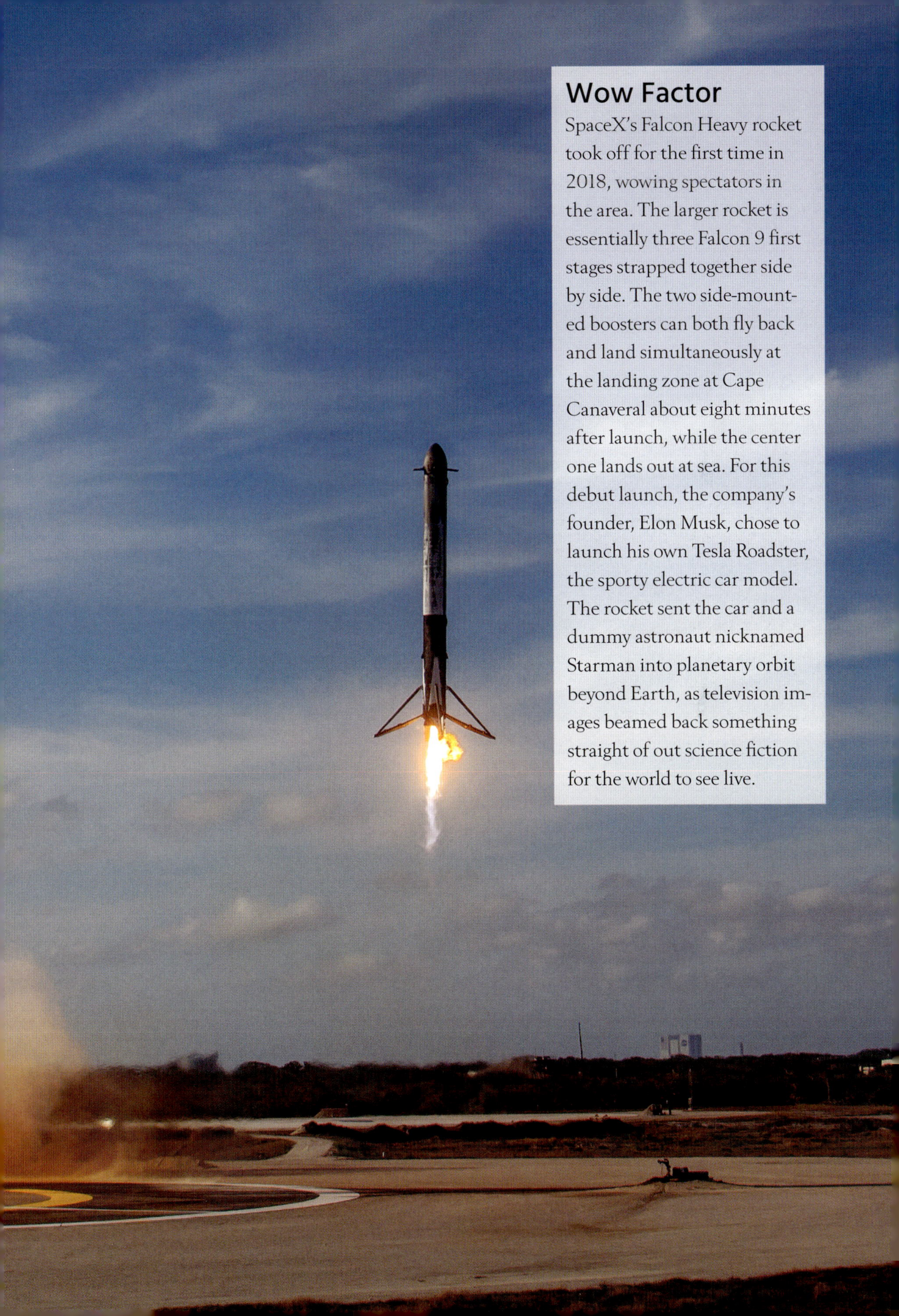

Wow Factor

SpaceX's Falcon Heavy rocket took off for the first time in 2018, wowing spectators in the area. The larger rocket is essentially three Falcon 9 first stages strapped together side by side. The two side-mounted boosters can both fly back and land simultaneously at the landing zone at Cape Canaveral about eight minutes after launch, while the center one lands out at sea. For this debut launch, the company's founder, Elon Musk, chose to launch his own Tesla Roadster, the sporty electric car model. The rocket sent the car and a dummy astronaut nicknamed Starman into planetary orbit beyond Earth, as television images beamed back something straight of out science fiction for the world to see live.

INDEX

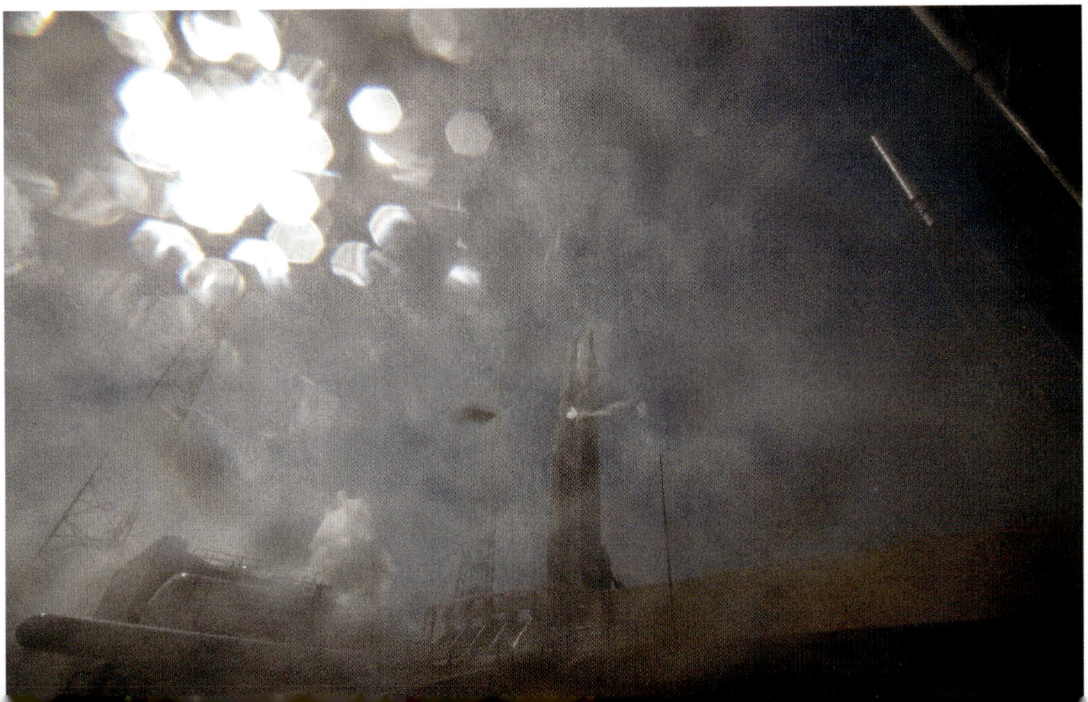

AmherstMedia.com

- *New books every month*
- *Books on all photography subjects and specialties*
- *Learn from leading experts in every field*
- *Buy with Amazon (amazon.com), Barnes & Noble (barnesandnoble.com), and Indiebound (indiebound.com)*
- *Follow us on social media at: facebook.com/AmherstMediaInc, twitter.com/AmherstMedia, or www.instagram.com/amherstmediaphotobooks*

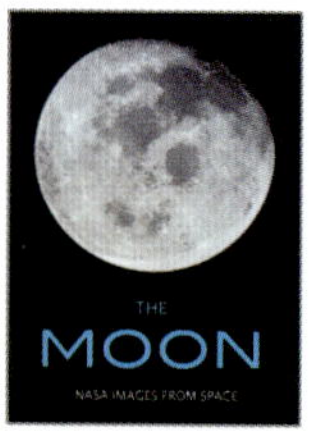

The Moon

NASA IMAGES FROM SPACE

Take a trip to our nearest neighbor in the night sky and explore the alluring beauty and scientific wonder of the moon. *$24.95 list, 7x10, 128p, 200 color images, index, order no. 2194.*

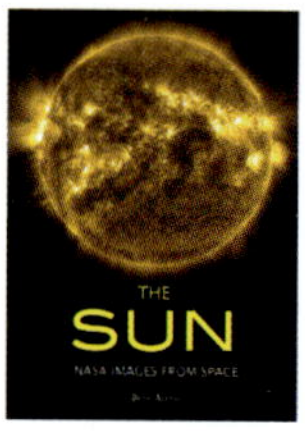

The Sun

IMAGES FROM SPACE

Travel to the center of our solar system and explore the awesome beauty of the star that fuels life on earth. *$24.95 list, 7x10, 128p, 180 color images, index, order no. 2180.*

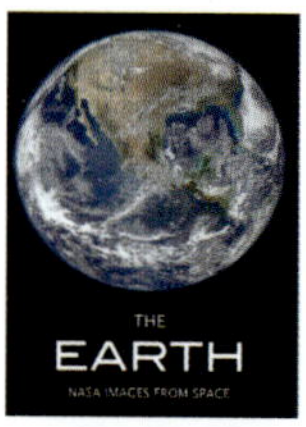

The Earth

NASA IMAGES FROM SPACE

Images from space reveal the startling beauty and incredible diversity of the blue orb we call home. *$24.95 list, 7x10, 128p, 180 color images, index, order no. 2170.*

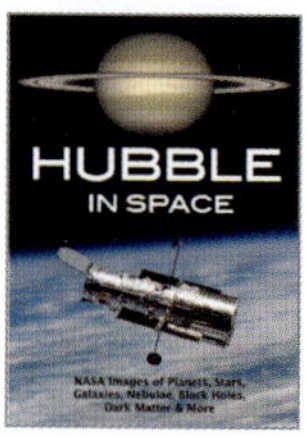

Hubble Images from Space

The Hubble Space Telescope launched in 1990 and has recorded some of the most detailed images of space ever captured. *$24.95 list, 7x10, 128p, 180 color images, index, order no. 2162.*

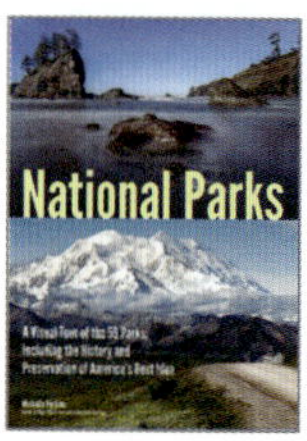

National Parks

Take a visual tour through all 59 of America's National Parks, exploring the incredible histories, habitats, and creatures these lands preserve. *$24.95 list, 7x10, 128p, 375 color images, index, order no. 2193.*

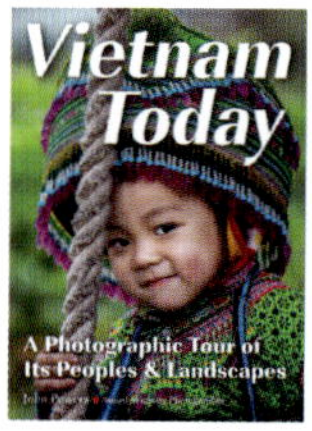

Vietnam Today

Vietnam veteran and acclaimed travel photographer John Powers explores both the rural and urban sides of Vietnam today. *$29.95 list, 7x10, 128p, 180 color images, index, order no. 2207.*

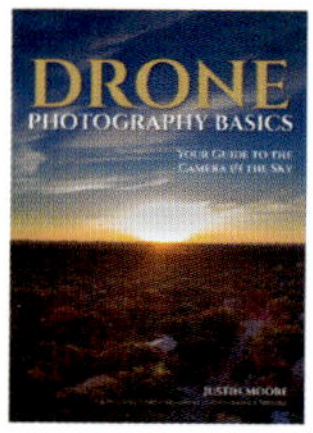

Drone Photography Basics

Justin Moore teaches you the fundamentals of using a drone and creating top-notch aerial photographs. *$34.95 list, 7x10, 128p, 130 color images, index, order no. 2211.*

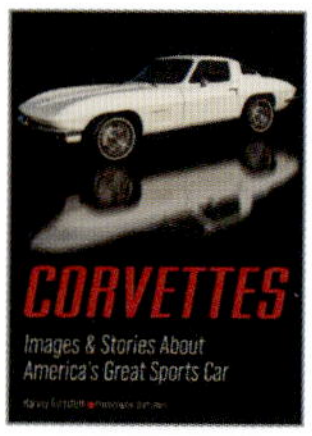

Corvettes IMAGES & STORIES ABOUT AMERICA'S GREATEST SPORTS CAR

Harvey Goldstein talks with owners from around the country about their love affair with these iconic cars. *$24.95 list, 7x10, 128p, 300 color images, index, order no. 2212.*

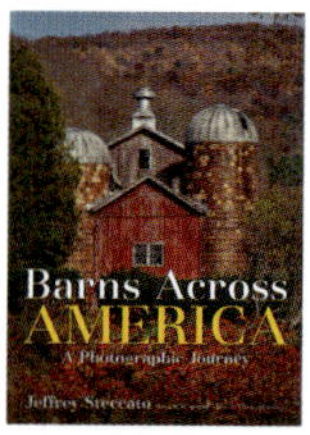

Barns Across America

A PHOTOGRAPHIC JOURNEY

Take a nostalgic look at America's great barns—some dilapidated, and others gloriously restored. *$29.95 list, 7x10, 128p, 160 color images, index, order no. 2214.*

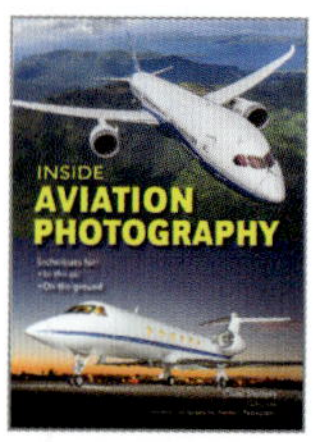

Inside Aviation Photography

Take wing with Chad Slattery as he photographs from the air and ground, shooting planes, pilots, and more. *$24.95 list, 7x10, 128p, 180 color images, index, order no. 2167.*